A study on Commercial and Industrial Electricity Consumers of Klang Valley, Malaysia

PERCEPTION OF ENERGY EXPERTS ON THE ADOPTION OF ENERGY EFFICIENT TECHNOLOGY

Ir. Dr. Thirumalaichelvam Subramaniam P.E., REEM

INDIA · SINGAPORE · MALAYSIA

Notion Press

Old No. 38, New No. 6
McNichols Road, Chetpet
Chennai - 600 031

First Published by Notion Press 2019
Copyright © Thirumalaichelvam Subramaniam 2019
All Rights Reserved.

ISBN 978-1-64546-799-1

This book has been published with all efforts taken to make the material error-free after the consent of the author. However, the author and the publisher do not assume and hereby disclaim any liability to any party for any loss, damage, or disruption caused by errors or omissions, whether such errors or omissions result from negligence, accident, or any other cause.

No part of this book may be used, reproduced in any manner whatsoever without written permission from the author, except in the case of brief quotations embodied in critical articles and reviews.

Dedication

I dedicate this book to my family, my wife, Evelyn Braok, who was invaluable throughout this process and in my life. Without her care, love, dedication, patience, and support, I would not have accomplished all the things I have done so far in my life. Evelyn, with untiring effort, has supported me in my career aspirations and has helped me achieve all my life goals by providing the necessary energy and emotional support when in times I was too tired to continue and was about to call it an end. To my daughter, Neesha Thirumalaichelvam, who, despite her own undergraduate and graduate studies, provided valuable advice and to complete this book. She has been instrumental in bringing out the best in me in order to complete this book by discussion points and making valuable suggestions as I had done for her own studies. To my son, Navien Thirumalaichelvam, who is in the final year of his undergraduate engineering degree, for supporting my research by teaching me various techniques in Microsoft Word and Excel. Both children gave me the extra push needed to complete this book in a timely manner. To my parents, Subramaniam and Parvathi, for inspiring me to fulfill my dreams. Your unconditional love, continued encouragement, and strong values have instilled unmentionable strength in the person I have become. Each time I fall, I am able to stand up and forge ahead because you have both taught me how not to give up. As a child, I saw how you were supportive of each person who sought your help. I, too, have now become that same person as I am now carrying forward what you have given. I thank you, Mom and Dad, for all of your sacrifices throughout my life. I am proud to carry on your legacy.

Perception of Energy Experts on Commercial and Industrial Electricity Consumers (CIEC) of the Klang Valley, Malaysia toward the Adoption of Energy-Efficient Technologies.

Ir. Dr. Thirumalaichelvam Subramaniam P.E., REEM

Abstract

Barriers to commercial and industrial energy efficiency improvements in Klang Valley, Malaysia are more pronounced due to the existence of factors such as weak policy and regulatory frameworks, economic and financial constraints, lack of information, and other issues. This research utilized a qualitative research methodology using a phenomenology approach aimed at enhancing the knowledge of commercial and industrial energy efficiency in Klang Valley, Malaysia by investigating the barriers associated with the implementation of energy efficiency measure. The eleven main themes and twenty-eight sub-themes identified from the study revealed that energy is poorly managed in the various commercial and industrial sectors and that there is an energy efficiency gap resulting from the low implementation of energy efficiency measures. In addition, the study revealed that the most important factors impeding the implementation of cost-effective energy efficiency technologies in the organizations are principally economic and financial barriers such as lack of budget funding and access to capital. The study also revealed that these economic and financial barriers are linked to the lack of adequate government framework for commercial and industrial energy efficiency. The study also showed that market factors related to cost reductions resulting from lowered energy use and threats of rising energy prices are the most important drivers for adapting energy efficiency technologies. To motivate energy efficiency, there should be established standards, guidelines, roadmaps, regulations, and enforcement of regulation suitable for the local environment, which at present has not been executed completely in Malaysia.

Keywords: Commercial and industrial, energy efficiency, Klang Valley, Malaysia.

Table of Contents

List of Figures   15

List of Tables   17

List of Abbreviations   19

List of Appendices   23

Acknowledgments   25

Chapter 1 Overview   27

Overview of Environmental Factors in Klang Valley, Malaysia   27

    Economy and Demographics   27

    Energy-Efficient Technologies in Malaysia   33

Research Problem   35

Theoretical Background   37

Purpose of this Research   39

Significance of Research   40

Research Questions   43

Assumptions and Limitations   43

Operational Definitions   46

Summary   48

Chapter 2 Literature Review   49

Research Paradigm Assumptions   49

    The Ontological Assumptions   49

    The Epistemological Assumptions   50

The Axiological Assumptions	50
The Rhetorical Assumptions	51
The Methodological Assumptions	52
Theoretical Orientation	52
Taxonomy of Barriers to Energy Efficiency	52
Armel's Research Barriers	53
Kadam's Research Barriers	54
Thollander et al.'s Research Barriers	55
This Study's Research Barriers	56
General Barrier in Implementing Energy Efficiency Technologies	59
EE Technology – Barrier	59
Determination of Barrier	61
Policy and Regulatory Barriers	63
Lack of a Legislative Framework	67
Economic, Financial, and Market Barriers	68
Behavioral, Information, and Technical Barrier	74
Implementation of Energy Efficiency Technologies	80
Existing Implemented Models on Energy Efficiency Technologies	80
The Use of Energy Control Systems	84
Standardizing Energy Management System	85
Reports of Energy Efficiency Technologies	88
Applied EE Metric	89
Critique of Previous Research	93
Usefulness of Implementing Energy-Efficient Technologies for the Consumers	93
Increased Gain on Markets	95

Increase in Energy Efficiency Technology Usage Leading to Economic Development	96
General Adaptability Related Model for the Adoption of Energy-Efficient Technologies	98
Core Competencies Framework	99
Camden's Behaviors Framework	100
Competency Framework	103
Motivation-Capability-Implementation-Results Framework (MCIR)	104
Grid Edge Actionable Framework	105
Summary	107
Chapter 3 Research Methodology	**108**
Research Design	108
Research Philosophy	110
Research Philosophy Stance	113
Research Approaches	116
Research Strategy	117
Research Choice	118
Research Techniques and Procedures	118
Selection Criteria	119
Selection of Participants	119
Research Questions	120
Population and Sampling Strategy	120
Research Instrument	121
Utility and Credibility of the Research	121
Role of the Researcher	122
Researcher Bias	122
Sources of Data (triangulation)	123

Data Collection Procedures	125
Data Collection Methods	125
Selection of Interview Questions	126
Conducting the Study	126
Secondary Data Collection	126
Data Analyses	127
Ethical Considerations	128
Scope and Delimitation of the Study	129
Summary	129
Chapter 4 Analysis and Presentation of Results	**131**
Demographic Statistics	132
Details of Analysis and Results	134
Presentation of Data	139
RQ#1: Resistance to Adopting Energy-Efficient Postures?	139
RQ#2: How Can Energy Efficiency Technologies be Implemented?	150
RQ#3: How Do Consumers find Energy-Efficient Technology Useful?	160
Summary of Themes and Major Findings	166
Summary	168
Chapter 5 Conclusion and Findings	**169**
Summary of the Results	169
Discussion of the Results	173
RQ#1: Resistant to Adopting Energy Efficiency Postures?	173
RQ#2: How Can Energy Efficiency Technologies be Implemented?	179
RQ#3: How Do Consumers Find Energy-Efficient Technology Useful?	184

Conclusion	191
Conceptual Framework	195
Non-Awareness of EE Management Techniques	197
Lack of EnMS Implementation	197
Lack of Government Support	197
Affordability of Policies	198
Lack of Information on EE Technologies	199
Financial Scheme Availability	199
Recommendations	200
Recommendation for Future Research	204
References	*207*

List of Figures

Figure 1	Location of Klang Valley region in Malaysia	30
Figure 2	Sub-regions of Klang Valley	30
Figure 3	Malaysian energy consumption by sectors from 1978–2015	34
Figure 4	Malaysian GDP from 1980–2015	35
Figure 5	Final energy demand by fuel type from years 1978–2015	41
Figure 6	Combined heat and power system: efficiencies	70
Figure 7	EnMS process	86
Figure 8	Core competencies framework	99
Figure 9	Camden's behaviors framework	101
Figure 10	Leadership competencies framework	103
Figure 11	Motivation-Capability-Implementation-Results (MCIR) framework	104
Figure 12	Grid Edge Actionable Framework	106
Figure 13	Saunders's Research Onion Methodology	109
Figure 14	Deductive process	116
Figure 15	Inductive Process	117
Figure 16	Data Source Triangulation	124
Figure 17	Summary of Academic and Professional Qualifications	134
Figure 18	Summary of Experience as Energy Expert	134
Figure 19	Saving results for air-condition No. 1	186
Figure 20	Savings results for air-condition No. 2	187

Figure 21	Combined savings results for air-condition No. 1 & No. 2	188
Figure 22	Savings by retrofitting to Nano LED Bulbs	189
Figure 23	The Research Conceptual Framework	196
Figure 24	GAT scoring card visualization matrix	201

List of Tables

Table 1	Barriers identified in Armel's research	53
Table 2	Barriers identified in Kadam's research	54
Table 3	Barriers identified in Thollander et al.'s research	55
Table 4	The main and sub-categorization of barriers identified based research reviewed	56
Table 5	Research design utilizing Saunders's "Research Onion Methodology"	110
Table 6	Participants Characteristics: Age and Gender	133
Table 7	Research Study Themes and Sub-Themes	136
Table 8	Participants' Responses: Theme #1	140
Table 9	Participants' Responses: Theme #2	142
Table 10	Participants' Responses: Theme #3	143
Table 11	Participants' Responses: Theme #4	145
Table 12	Participants' Responses: Theme #5	147
Table 13	Participants' Responses: Theme #6	150
Table 14	Participants' Responses: Theme #7	153
Table 15	Participants' Responses: Theme #8	154
Table 16	Participants' Responses: Theme #9	157
Table 17	Participants' Responses: Theme #10	161
Table 18	Participants' Responses: Theme #11	163
Table 19	Summary of Nano Led Lighting Retrofitting Project	189

List of Abbreviations

AEMAS	ASEAN Energy Management Scheme.
AEPS	Alternative Energy Portfolio Standard.
BAS	Building Automation System.
BATs	Best Available Technologies.
BEE	Bureau of Energy Efficiency.
BEMS	Building Energy Management System.
BIMS	Building Information Management System.
BIS	Building Intelligent System.
CDM	Clean Development Mechanism.
CEOs	Chief Executive Officers.
CEPS	Clean Energy Portfolio Standard.
CHP	Combined Heat and Power.
CIEC	Commercial and Industrial Electricity Consumers.
CIS	Commonwealth of Independent States.
CMMS	Computerized Maintenance Management System.
CO_2	Carbon Dioxide.
CSD	Constant Speed Drive.
CSR	Corporate Social Responsibility.
DES	District Energy Systems.
DOE	Department of Energy.
DOST	Department of Science and Technology.

EBED	Energy-Based Economic Development.
ECBC	Energy Conservation Building Code.
EE	Energy Efficient or Energy Efficiency.
EEAP	Energy Efficiency Action Plan.
EECAM	Energy Efficiency and Conservation Agency Malaysia.
EELP	Energy Efficiency Leading Programme.
EERS	Energy Efficiency Resource Standard.
EnMS	Energy Management System.
ESCOs	Energy Services Companies.
EU	European Union.
G20	Group of 20 Nations that Represents 85% of Global Economies.
GBI	Green Building Index.
GDP	Gross Domestic Product.
GHG	Greenhouse Gases.
$GtCO_2$	Billion tonnes of Carbon Dioxide.
GWh	Gigawatt Hour.
HVAC	Heating, Ventilation and Air-Conditioning.
IRR	Internal Rate of Return.
ISO	International Standardization Organization.
km^2	Kilometer Square.
Ktoe	Kilotonnes of Oil Equivalent.
kWh	Kilowatt Hour.
LED	Light Emitting Diode.
LEED	Leadership in Energy and Environmental Design.
m^2	Meter Squared.

MBA	Master of Business Administration.
MCIR	Motivation Capability Implementation Framework.
MEPA	Malaysia Energy Professional Association.
MEPS	Minimum Energy Performance Standards.
Mt	Megatonnes.
MWh	Megawatt Hour.
NAMA	Nationally Appropriate Mitigating Action.
NEEAP	National Energy Efficiency Action Plan.
NEEMP	National Energy Efficiency Master Plan.
NGO	Non-Governmental Organizations.
NPV	Net Present Value.
O&M	Operation and Maintenance.
OECD	Organization for Economic Co-operation and Development (OECD).
OFDMA	Orthogonal Frequency Division Multiple Access.
PE	Professional Engineer.
PhD	Doctor of Philosophy.
PREE	Peer Review Energy Efficiency.
QMS	Quality Management System.
R&D	Research & Development.
RDD&D	Research, Development, Demonstration, and Deployment.
RE	Renewable Energy.
REEM	Registered Electrical Energy Manager.
ROI	Return on Investment.
RPS	Renewable Portfolio Standards.
SEAD	Super-efficient Equipment and Appliances Deployment.

tCO_2	Metric Ton Carbon Dioxide.
tCO_2e	Metric Ton Carbon Dioxide Equivalent.
TOP TENs	Best Available Technologies and Practices.
UNIDO	United Nations Industrial Development Organization.
VSD	Variable Speed Drive.

List of Appendices

Appendix A – Consent Letter   235

Appendix B – Individual Consent   237

Appendix C – Interview Protocol   239

Appendix D – Interviewee's Transcript Review (ITR) Letter   246

Acknowledgments

First of all, I give thanks to the Lord for guiding me and making it possible for me to complete my doctoral studies despite numerous challenges. God helps me in mysterious ways, often through those who are close to me. My parents, Subramaniam and Parvathi, my brother, Thamizhchelvam, and my sister, Dr. Thamizhchelvi, were my building blocks who helped create the individual I am today. Even though there were many challenges while conducting the research for this book, it helped widen my horizons, and I viewed the world in a different perspective while enriching my experience in which I owe my gratitude to everyone who supported, inspired, and guided me.

Secondly, I would like to extend my gratitude to the SMC University. I would also like to thank the professors in Faculty of Management, all of whom aided in contributing to my knowledge. I would like to especially thank Dr. Albert Widman, Dr. Louis F. Carfagno, Dr. Thomas Grisham, Dr. Nikhil Agarwal, Dr. Sureneetha Harath, and Dr. Mysoon A. Otoum. In addition, I sincerely thank all members of the SMC faculty, in particular, Jessica Harlan Phipps, for their support and advice to enable me to accomplish this book.

Dr. Jeffrey Ray very kindly obliged me by accepting to be my mentor and provided tireless assistance by attending to chapter drafts of this book, fielding questions and concerns, providing sound advice, constructive comments, continuous encouragement, and moral support when I was confused with certain theoretical concepts. He always had time for me by supporting me at inconvenient times to address urgent issues. Dr. Ray, with his vast knowledge and experience, had further fueled my interest in this subject and brought out the best in me by guiding me enthusiastically throughout the research and writing of this book in order to reach my objectives. It would have been impossible for me to succeed without his guidance and support.

I would also like to thank Dr. Ted Sun, Vice Chancellor of SMC University who went out of his way to support me over e-mail with excellent guidance to my many inquiries. With his warm encouragement, he provided me with a strong motivation to complete this book.

I am forever grateful to my sister Dr. Thamizhchelvi Subramaniam, Dean of Student Success at Compton College, California, USA, who has continuously provided her support not only in my life but by reviewing and proofreading my book.

Finally, I would like to show my appreciation by thanking all participants for taking time from their busy schedules to volunteer in this research and for providing their perceptions and views from their vast experiences in order to complete this book in an open and professional manner.

Chapter 1

Overview

Since Malaysia's independence in 1957, its standard of living has improved largely due to an economic shift from agriculture to industrialization. According to the report of the World Development Indicator by the World Bank (2016), Malaysia's economy is noticeably boosted for further growth. The rise in ownership of electrical appliances in all major sectors has led to an increase in energy consumption, which has consequently affected the environment. This research aims to encourage the adaptation of energy-efficient (EE) technologies from the perspectives of energy experts of Commercial and Industrial Electricity Consumers (CIEC) in Klang Valley, Malaysia (World Bank, 2016). This research uses a qualitative research method with a phenomenological technique to study the lived experiences of ten energy experts. The primary data is collected through efforts done by the researcher. Primary data was collected through, interviews, diaries, focus group, observations, portfolios, incidents, and personally collected information by the researcher. Secondary data was collected from information released by the government, organizations, departments, records, and data from previous research. An interview was conducted where the information acquired provided insight into EE technologies implementation. After collecting primary data from the energy experts, secondary data was used to support the primary data. The data collected provided answers and themes to research questions.

OVERVIEW OF ENVIRONMENTAL FACTORS IN KLANG VALLEY, MALAYSIA

Economy and Demographics. Malaysia's economy is dependent on its manufacturing and exporting of electronic goods (Thomas White International, 2014). The energy used for manufacturing these electronic goods is generally high. Hence, in order to save energy, other resources or products that can be exported, such as natural products, are needed.

Examples of these resources or products are live animal stock, animal meat, fishes, dairy products, life plants, coffee, tea, cereals, edible fruits, and vegetables. Exporting natural resources rather than utilizing natural resources could be a solution to reducing or minimizing energy usage for exported items, which could have increased profits even if the availability of natural resources were limited (Tan & Nikkei, 2017). This was essentially the case during the economic crisis of 2009, when the production and distribution of natural resources such as palm oil had decreased. In addition to the decline in production and distribution of palm oil, the production of crude rubber (latex) to manufacture tires and other rubber (latex) products for export to China and India took the same fate as palm oil during this economic crisis as Malaysia was less reliant on exporting natural resources to generate revenue (Lopez, 2016). Due to these reasons, proposals to export other resources or products did not have priority at this time. Thus, there continues to be a high demand for energy use to manufacture and export electronic goods to boost Malaysia's economy (Sadler, 2017).

Malaysia's economy relies on the many opportunities from the trade agreements it has with Hong Kong and Singapore. According to the Asia Development Bank (ABD) (2011), Malaysia, when compared to many other developing countries, is on its way to becoming an economically developed nation when considering the level of improvement it has undergone according to reports of previous papers. Hence, Malaysia is a developing country, its citizens are considered "middle-income families," and it has a balanced economy compared to other countries. Malaysia's strategy to balance its economy is to reduce poverty by utilizing all possible opportunities and to improve trade efficiency (Hezri, 2016).

From the annual reports of economic status by Koen, Asada, Nixon, Habeeb Rahman, and Mohd Arif (2017), Malaysia had many economic crises that were solved by enacting laws and guidelines in order to mitigate the crises. In 2015, the Malaysian economy grew by five percent due to the planned management of monetary laws. The Monetary Policy Committee provided a plan to ensure Malaysia's sustainable development for the year 2015, a time when economic overheating was at its peak and when the threat of inflation was high. The strategy was to tighten the monetary policy by withdrawing funds from the banking system and raising interest rates. The higher interest rates would encourage people to save more and spend less. It would also make it more expensive for people to borrow money. The plan then caused consumption and investment to slow down to a level that was more sustainable and reduced the prospect for high inflation. Conversely, when economic conditions weakened, funds would be injected into the banking system to reduce interest rates and spending and borrowing would increase. The resulting increase in consumption and investment would stimulate further economic activity, leading to higher income, employment, and economic growth. The implementation of this act contributed to a growth of about half a percent greater than the previous year. The monetary plan actually targeted both residential and industrial related growth; however, the development and benefits were generated from governmental industrial sectors rather than any private organization's contribution (Bank Negara Malaysia, 2015).

Malaysia's external debt is a subtle task for the government to manage, but the implemented monetary development laws helped manage the external debts crisis for several years. Malaysia's external debts, with the aid of contributions from within its governmental organization, was managed according to its annual report (Bank Negara Malaysia, 2015). According to the report by the Bank Negara Malaysia (2015), the external debts that were incurred by the private sectors through overseas business dealings would affect Malaysia's debts if the duration of the business projects were greater than two years.

The economy is influenced by the demographics and its population. Figure 1 shows the location of Klang Valley in Malaysia, and Figure 2 shows the sub-regions of Klang Valley comprising of six districts, which are Petaling Jaya, Gombak, Klang, Kuala Selangor, Kuala Langat, and Ulu Langat. This region which has a population of 7.2 million over an area of 243 km^2 (kilometer squared) (Department of Statistics Malaysia, 2017). The ethnic composition of its people is 50.1% Malay, 22.6% Chinese, 11.8% Orang Asli (aboriginal people), 6.7% Indian, 0.7% others, and 8.2% non-citizens. The religious makeup of people in the Klang Valley is 61.3% Muslim, Buddhist 19.8%, Christian 9.2%, Hindu 6.3%, Confucianism, Taoism, other traditional Chinese religions consist of 1.3%, others 0.4%, none 0.8%, and unspecified 1% (Index Mundi, 2016). The Klang Valley accounts for 20 percent of Malaysia's population and it adds RM263 billion to the gross national income in 2010 (Inside Investor, 2012). With over 1,600 foreign companies having offices in the Klang Valley, it provides an international hub of business to the local market, which contributes significantly to the local Gross Domestic Product (GDP) per capita which has risen presently to US$10,060 from US$7,059 in 2009 (Malay Mail Online, 2017). Malaysia is ranked 18th overall among the 189 economies covered in the survey conducted by the World Bank (World Bank Group, 2016).

Figure 1 Location of Klang Valley region in Malaysia. Adapted from "INTA International Urban Development Association," by inta-aivn (2017).

Figure 2 Sub-regions of Klang Valley. Adapted from "Department of Statistics Malaysia" (2017).

The Klang Valley metropolis region is faced with issues arising from urbanization and the need to attract more investments. When comparing other progressive countries in the 21st century, Malaysia has the most effective economic strategy; however, the rate of development is lower when compared to other developing countries. Indonesia, Philippines, Singapore, and Thailand are the top attractive places for foreign investment, surpassing Malaysia. Hence, in order to sustain Malaysia's economy, there is a need to implement a green environment. Implementation of the green environment directly affects the growth of Malaysia's economy and increases foreign investments, and moreover, interstate migration plays an important role in achieving increased foreign investments (Abdullah, Abu Bakar, Mohd Jali, & Ibrahim, 2017). A survey conducted by the Malaysian government has noted that there are a number of internal policies within the country that were implemented for economic balance. The survey states that current development has increased by about half a percentage of the previous year's development. Moreover, the survey adds that these internal policies act as a balance among various factors that influence the economy, thus paving the way for increased use of EE technologies. The survey further adds that the growth for implementing EE technology has increased by about 2.6 percent in comparison to the expected population increase (Tey, 2014).

The macro environment of the Klang Valley is in the transition from middle to upper class as families are earning more. With the new earned income, these families are now enjoying newer and improved technologies which use more energy. The need to support the increased energy use requires a simultaneous growth in communication and infrastructure. The population has an indirect effect on the environment. With more people moving from middle to upper class, their lifestyle too changes, causing an unsustainable growth in private transport demand. This problem is made worse by the declining public transport modal share from 34% in 1985 to 20% in 1997 and a further drop to 18% in 2009 (ERE Consulting Group, 2017). With the increase in private transportation and a decline in public mode of transportation, there is an increase in pollution, which could have a detrimental effect on the environment. Additionally, insufficient infrastructure within an organization affects the quality of equipment, and the amount of harmful emissions from this additional equipment is high (RVO, 2017). Even though raw materials are decreasing in the Klang region, there is a need to provide required raw material resources to sustain the population's needs, thus further increasing the demand for transportation. Demands of infrastructural necessities have to aim for a sustainable environment by utilizing efficient technology (RVO, 2017). The increased number of road systems and the use of excessive transportation make the increased consumption of energy inevitable.

Even though Malaysia is active in international trade and business, the country still faces many challenges in implementing EE systems. About 51% of the population is employed in manufacturing industries, and the number of professionals and immigrants are increasing daily as the economy improves. The 7.2 million people in this area affect 40% of the GDP of Malaysia; however, the utilization of green technologies is lacking (RVO, 2017). Compared to other neighboring countries such as Indonesia, Singapore, Philippines, Brunei, Thailand, and Vietnam, Malaysia's implementation of energy efficiency is approximately 34% more. The Malaysian government has invested more on smart grids for efficient energy usage. This includes transportation usage which has increased; however, the advancement on roadways is not pedestrian-friendly since the curbside walkways are not retrofitted for handicap access (RVO, 2017). People from other countries come to reside in Malaysia for various purposes such as work, business, education, and leisure, which have further added to the demand for energy. Efforts to reduce energy demands by recycling used material turned out to be only 11% compared to that of Denmark and Singapore (PEMANDU, 2012). Observatorio Asia Pacifico (2016) reported that the Eleventh Malaysia Plan (2016–2020) states its three-dimension targets to be the environment, people, and the economy. If the Eleventh Malaysia Plan was to be fully implemented, then the welfare for sustained environmental growth, human capital, and infrastructure support changes, as wasteful and non-productive patterns in energy development ceases. Malaysia's National Energy Efficiency Action Plan (NEEAP) concentrates on achieving its objectives by targeting five-star energy appliances with the help of government, commercial, and industrial sectors. Implementation of the International Standardization Organization (ISO) 50001 Energy Management standards in enhancing EE buildings and cogeneration is one of its ultimate targets for commercial buildings. Its target is to attain commercial building electricity consumption to be in the region of 200–300 kWh/m^2/year. The benefits of this implementation indirectly influence economic growth in terms of energy savings. One unit of energy saved in the demand-side saves three to four units of primary fuels. In addition, investments in energy supply facilities such as power plants and electricity transmission grid system can be deferred or postponed (KeTTHA, 2014).

Energy-Efficient Technologies in Malaysia. Most building sectors have the potential to implement EE technologies, which helps reduce Greenhouse Gases (GHG) emissions. Many building sectors have already taken steps to implement EE based on current knowledge available in EE technologies. Hence, building sectors have already taken steps to implement roles for an EE future. Saving one unit of energy at the demand-side can save primary energy by three to four times if effective energy technologies are utilized (KeTTHA, 2014). In an effort to promote the use of effective energy, the Malaysian government provided tax incentives for the use of EE technologies as a means to address the issues related to the lack of professionals capable of implementing these technologies (Eang, 2015). Despite the Renewable Energy (RE) plan implemented in Malaysia, it could not reach the goals of a sustainable environment even with the availability of many alternate sources such as biogas, biomass, mini-hydro, and solar energy. Depletion of indigenous energy resources and mass climate changes has compelled the Malaysian government to embrace and implement energy efficiency policies. The move to implement energy efficiency policies by the Malaysian government was to reduce energy consuming sources required in order to boost the economy. The Peer Review on Energy Efficiency (PREE) team has worked with the Malaysian government to implement energy efficiency. The actions of the team are aimed at promoting an effective EE framework for improving energy efficiency use (APEC, 2011). Malaysia is embracing the promotion of EE designation by strengthening institutional capability known as Energy Efficiency and Conservation Agency Malaysia (EECAM). The National Energy Efficiency Master Plan (NEEMP) was implemented for three main economic sectors from the period of 2011 through 2020. These sectors include industrial, commercial, and residential. The transportation industry is the second largest energy consumer taking up to 36.9% of the energy resources of Malaysia. Many programs were introduced to indirectly reduce energy use such as green vehicles and to make it compulsory for the population to adapt to green energy technology (APEC, 2011).

From the governmental side, there were many actions taken on residential, commercial, and industrial sectors. In 2014, incandescent light bulbs were phased out in residential sectors. The government highlighted the increased use of energy labeled products for personal use. Minimum Energy Performance Standards (MEPS) was implemented for appliances introduced to households (Energy Commission of Malaysia, 2014).

According to the Malaysia Energy Information Hub (2016), the transport sector of Malaysia was the largest energy user, accounting for 45.5% of the total energy demand in 2015. The industrial sector was ranked at 27%, the commercial and residential sectors with a share of 14.59%, followed by non-energy with 11.44%, and finally agriculture with 1.73%. Figures 3 and 4 illustrate the Malaysian industrial sector's demand for energy in kilotonnes of oil equivalent (Ktoe) from 1978 to 2015. From 2007 to 2015, the demand for energy has gradually decreased by an average rate of 1.68% per year, and its GDP had increased by an average rate of 4.65% per year.

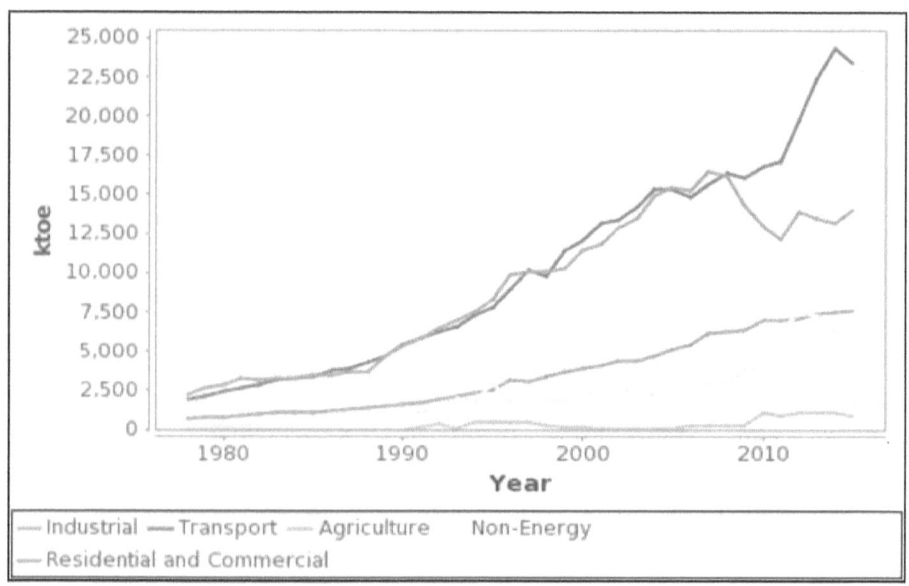

Figure 3 Malaysian energy consumption by sectors from 1978–2015. Adapted from "Introduction to Malaysia Energy Information Hub," by Meih (2017).

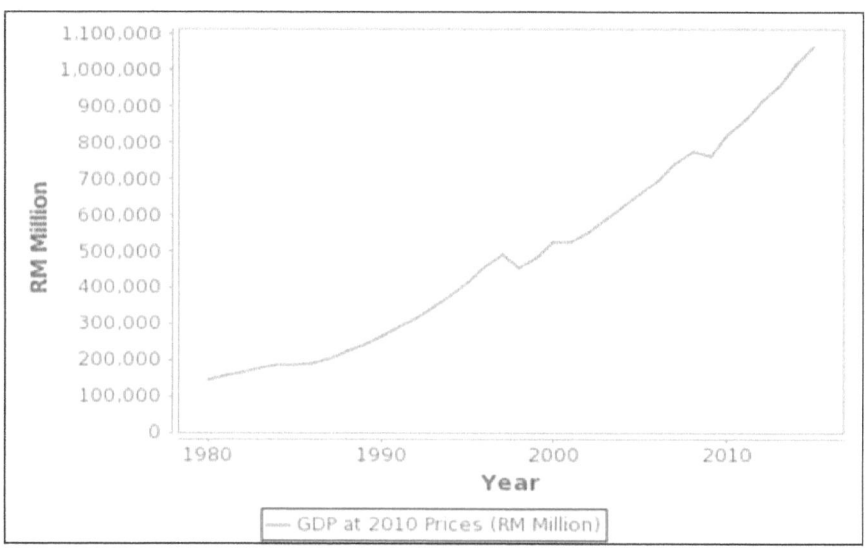

Figure 4 Malaysian GDP from 1980–2015. Adapted from "Introduction to Malaysia Energy Information Hub," by MEIH (2017).

These statistical results clearly indicate that there was an increased use of EE equipment and technologies in the industrial sector that brought about energy reduction during this period. With help from the government, Green Building Index (GBI) was developed to create a greater effect on the green economy in order to specifically aid the third largest consumers, the residential and commercial sectors, to reduce energy (KeTTHA, 2017).

RESEARCH PROBLEM

Azlina, Abdullah, Kamaludin, and Radam (2015) claim that industrial and commercial sectors in Malaysia use excessive electricity due to the usage of inefficient equipment consuming a tremendous amount of energy. Current trends indicate that there is demand in utilizing EE technology for appliances that consume high power (Menezes, Cripps, Buswell, Wright, & Bouchlaghem, 2014). According to DOE, the usage of inefficient equipment leads to increases in the cost of energy-related projects, which ultimately narrow profit margins by 20 to 35% (DOE, 2015). Since industrial and commercial sectors are using excessive amounts of energy, they are cutting their profits to pay for their use of energy. This decrease in profit of as much as 20 to 35% reduces reinvestments for other developmental activities, which may ultimately lead to non-implementation of additional projects (DOE, 2015b).

Increases in energy consumption threaten the ozone layer, leading to the emission of carbon dioxide (CO_2), which produces greenhouse effects. Between 1990 and 2004, Malaysia's carbon emissions grew by 221 percent (+221%) due to increased energy demand from residential, commercial, industrial, and transportation sectors. During this time period, Malaysia had the fastest economic growth rate in the world (Zaid, Myeda, Mahyuddin, & Sulaiman, 2014). Malaysia's carbon emission in 2014 was 185 million tonnes, which is approximately 0.6% of the global total (30, 276 million tonnes of CO_2). If the trend continues, Malaysia's CO_2 emission in 2020 amounts to 285.73 million tonnes, which is a 68.86% increase compared to the year 2000. Due to these serious problems faced with CO_2 emission due to increased energy demand, Malaysia voluntarily committed to reducing 40% of its Greenhouse Gas (GHG) emissions by the year 2020 as was announced at the 2009 United Nations Climate Change Conference in Copenhagen (COP-15) (Zaid, Myeda, Mahyuddin, & Sulaiman, 2014).

Developing countries such as Malaysia tend to construct more buildings for their residential, commercial, and industrial sectors, leading to various demands from transportation to electricity generation, resulting in an increased demand for energy (Hubacek, Guan, & Barua, 2007). To generate electricity from primary energy sources, Malaysia utilizes two major types of power generating plants, thermal and hydro. Thermal power plants transform primary energy sources like natural gas, coal, fuel oil, and biomass into electricity, whereas hydropower plants convert water-heads into electricity through water turbines. Among the total electricity generation of 134 gigawatt hours (GWh) in Malaysia, thermal stations have contributed the most, nearly 119 terawatt hours (TWh), whereas hydropower plants generated 9.0 TWh in 2013 (Energy Commission of Malaysia, 2014, 2016). Reserves of chief primary energy sources in Malaysia are 1.94 billion tons, including 15800 TWh of coal, and 0.64 billion tons that is 7700 TWh of crude oil, and 2784.1 billion standards cubic meters that is 31000 TWh of natural gas. Yearly production rate per day for crude oil in 2013 was 73.5 thousand cubic meters and 190.6 million cubic meters of natural gas. This data is important since it shows that fossil fuels, like natural gas, coal, and coke, have contributed most of the electricity generation, but their reserves are depleting very quickly (Samsudin, Rahman, & Wahid, 2016).

Hence, due to the problems brought about by the increase in energy use, the inefficient use of energy, and reserves of fossil fuel depleting very quickly, this research aims to study the perceptions of energy experts regarding commercial and industrial electricity consumers and discover the behavioral factors that affect adoption and utilization of EE technologies, particularly in the Klang Valley region of Malaysia. The current energy reduction awareness programs being adopted could be improved and strengthened by utilizing the behavioral factors that will be discovered in this research (KeTTHA, 2014).

THEORETICAL BACKGROUND

Materials needed for energy efficiency implementation have been abundant since Malaysia's independence; however, Malaysia tends to depend on alternative resources for its growth. The government has put in place adequate initiatives for sustainable development by ensuring maximum utilization and minimal reduction of waste. About a 6.3 percent increase in energy demand is expected annually, exceeding the total value of GDP, which is expected to grow at 6.0 percent annually (KeTTHA, 2014). According to Irrek and Thomas (2008), there are many types of energy efficiency usage that can be classified and given priority such as end-use energies, useful energies, energy end-use efficiency, and on-demand energies. End-use energy can be termed as the usage of energy by consumers after the production of a product such as long-distance heat, cooler, electricity, boiler, natural gas, and petrol. Useful energy is defined as the energy used during production such as radio-thermal heat, process heat, compressed air, kinetic energy, and light. On-demand energy can be defined as the process of using the energy for comfort such as refrigerators, cooling of living rooms, using heat for food preparation, and vehicles for personal transportation. Implementing EE technologies needs to be carried out for the future of humankind and not merely to fulfill a law or an act, but to help address the rapid degradation of our environment. The study by Khairunnisa, Yusof, Salleh, and Leman (2015), derived a framework for applying EE technologies in residential environments such as introducing certifying standards, act, law, appliances for production systems, and appliances scheduling. Theoretically, there are more barriers found in implementing EE technologies that are related to technical, financial, policy, and regulatory which can be addressed if governments take adequate action regarding these issues (UNIDO, 2017).

The majority of energy is expected to be consumed by manufacturing sectors such as in the iron, steel, cement, wood, gas, pulp, paper, rubber (latex), and ceramics industries (APEC, 2011). Coal is one of the primary fuel sources used for the generation of electricity in Malaysia. This coal generated electrical energy is mainly utilized by the iron, steel, and cement manufacturers which produce pollutants that have a detrimental effect on the environment due to their higher level of CO_2 and sulfur emission (APEC, 2011). The implementation of the Nuclear Power Infrastructure Development Plan and the Nuclear Power Regulatory Infrastructure Development Plan by the government would provide important steps toward developing nuclear power for Malaysia's future electricity supply. These support the multiple goals of improving energy security, spurring economic development, as well as reducing CO_2 and GHG emission. A new independent atomic energy regulatory commission could be established to regulate the civilian use of by-product, source, and special nuclear materials to ensure adequate protection of public health and safety, to promote the common defense and security, and to protect the environment. A 10-Year Comprehensive Communication Plan and Strategies on Nuclear Power for electricity continues to increase awareness and public acceptance (Zulkifli, 2016).

Decision makers in organizations base their decisions on the financial resources that are available to invest in EE projects. The decision makers are the best persons to determine if a particular technology is needed in their organizations based on the project's return on investment. If there is a lack of technical information on new EE technology products, there need to be steps implemented by authorities to ensure that the public has access to the information or the technology. Most energy consuming practices rely on sophisticated devices that are actually based on domestic and industrial appliances. Observing and understanding demonstrations of EE technology itself provide knowledge on how to implement these technologies as a start (EPA, 2009). The overall benefits of implementing EE technologies are to preserve good health, increase employment, boost the value of buildings or facilities, instigate industrial competitiveness, and to increase energy security for the future (Cambridge Econometrics, 2015). As mentioned earlier, there are four types of energy usage can be classified: end-use energies, useful energies, energy end-use efficiency, and on-demand energies. For example, if there is a need to produce EE equipment, then the manufacturer of the equipment should make use of EE technology where it can use combined energy resources such as gas or electrical power, e.g. hybrid-powered equipment. Hence, it is important to classify how energy is used. If the end-user's intention is to use energy for luxury, it can be avoided or modified accordingly. If the end-user does not mean to use energy intensively, then the energy needed for producing the products is not detrimental in the long run. Over 42 types of possible barriers were identified that affected energy efficiency across these four groups (DOE, 2015a).

On defining sustainable energy consumption, the ratio of the output energy must comply with the input of the energy used for production (Cambridge Econometrics, 2015). Manufacturing and construction companies are the main consumers of energy; hence, certain technologies can be applied to target energy efficiency. The major issues that influence the owner of the construction business to implement energy efficiency are the effects of wage on construction businesses, skills, work cultures, social cultures, unionization, collective actions, employee turnover, absenteeism, health, and supply. The government should provide incentives or implement laws which could help resolve these barriers which in turn encourages businesses to implement EE technologies (KeTTHA, 2014). From a worldwide perspective, countries with dense populations such as Argentina, Eritrea, Bangladesh, Myanmar, and Nepal, for example, use the least amount of energy since the demand and need for energy is low (Argentina Country Reports, 2011). Even though the Klang Valley is a smaller region in terms of land area, it is still as dense as the other densely populated countries mentioned earlier, but the energy usage is higher than these countries. Hence, this research intends to provide behavioral characteristics of CIEC toward the adaptation of EE technologies from the perspective of energy experts.

PURPOSE OF THIS RESEARCH

The primary purpose of this research is to study the perception of energy experts on CIEC of the Klang Valley toward the adoption of EE technologies. This research follows a qualitative research methodology, utilizing phenomenological techniques in capturing the perspectives of energy experts in the above study.

The Klang Valley is a major and densely populated region of Malaysia. The developments in this region are very rapid in comparison to the other regions of Malaysia. There are many issues pertaining to energy demand due to the increase in population. With the increase in population, there is an increase in infrastructures such as residential houses, commercial buildings, industrial facilities, and transportation, which directly increases energy demands. The major purpose of this research is to understand the behaviors and the common attitudes that are observed within the consumers of electricity in adapting energy efficiency usage from the perspective of the energy experts. Previous research has identified adaptability as one of the major barriers to energy efficiency usage. It is found that environmental, technological, policy, economical, regulatory, and investment are the other barriers (Backman, 2017; Chpassociation, 2015; Sussex, 2017).

Most research has been conducted on implementing EE technologies (Energy Community, 2017; EPA, 2009; IEA, 2011; OEE, 2012; Worrell et al., 2003). Research has also been conducted on policies implemented in Malaysia. Still, little is known as to what affects consumer's intentions and behavior in adapting EE technologies. Hence, this research aims to discover what the possible barriers are in implementing EE technologies. The research gives a clear roadmap on how the barriers can be properly identified and how the barriers can affect the decisions on the implementation of EE technologies. Open-ended questions are used to interview energy experts and data collected from the interviews provides solutions to address these issues. By the end of this research, specific information is compiled addressing the research objectives as a whole. This research also aims to understand the primary behaviors preventing the adaptability of EE technologies in the Klang Valley region.

SIGNIFICANCE OF RESEARCH

This research will identify why CIEC of Klang Valley, Malaysia are restricted from adopting available EE technologies. Since energy experts understand the behavior and attitudes of their customers, they will be in a better position to transfer information to the upcoming researchers, scholars, approximately one thousand energy experts (Energy Commission of Malaysia, 2018), and four authorities that are promoting EE technologies.

From these sources, this research will develop a solution on how EE technologies will be used and applied. This research will provide a solution based on the collected data with respect to how EE technologies can be implemented.

Malaysia's major drivers for the increased demand for energy are its population growth and economic expansion, mainly due to its manufacturing sector. The nation faces an imbalance in its ratio between energy demand and GDP, which is caused by its energy demand growth rates being higher than its GDP growth rates. This imbalance was due to the acceleration of its industrialization process over the past two decades (KeTTHA, 2014). The need to promote efficient use of energy has become clear, and this research study will contribute to sound energy efficiency policies, effective awareness, and programs that can be implemented to help overcome the imbalance mentioned earlier.

Figure 4 illustrates Malaysia's final energy demand by fuel types in Ktoe from the years 1978 to 2015. According to the Malaysia Energy Information Hub (2016), in 2015 motor petrol was the largest final energy demand by fuel accounting for 24.71% of fuel used. Electricity ranked second with a share of 21.96%, then natural gas with a share of 18.46%, followed by diesel with a share of 18.10%.

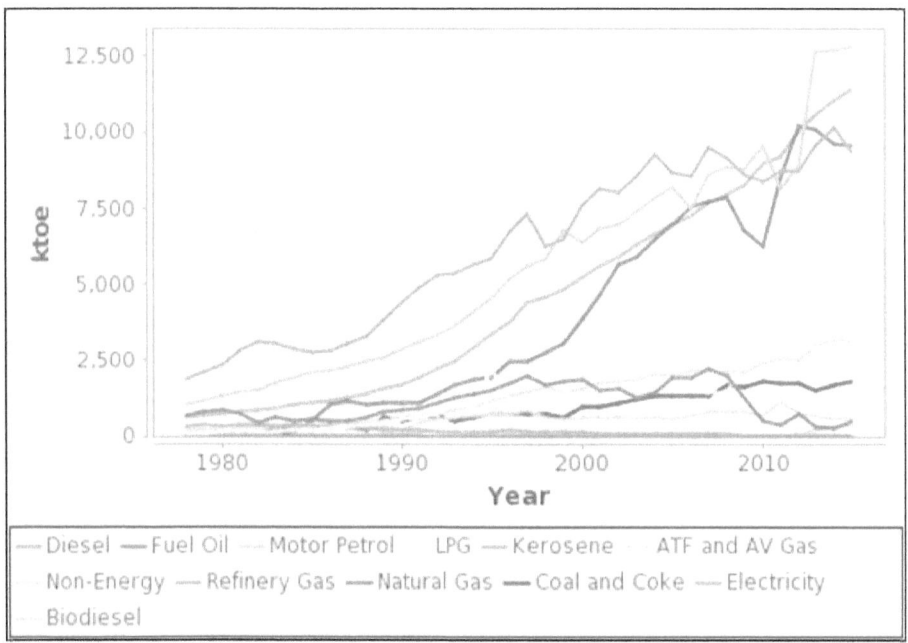

Figure 5 Final energy demand by fuel type from years 1978–2015. Adapted from "Introduction to Malaysia Energy Information Hub," by MEIH (2017).

The total electricity generated by Malaysia in 2015 was 144,565 GWh compared to the total electricity generated in 2005, which was only 94,030 GWh. This signifies a rapid increase of 53.74% in electricity generation within a short time span of ten years (MEIH, 2016). The overall average CO_2 emission per megawatt hours (MWh) of electricity production for Peninsular Malaysia, Sabah, and Wilayah Persekutuan Labuan was 0.615 Metric Ton carbon dioxide (tCO_2)/ megawatt hour (MWh) in 2014. The electricity generated by Malaysia in 2015 contributed to a CO_2 emission of 88.9 megatonnes (Mt). The total Global CO_2 emission for 2014 was 35.7 billion tonnes CO_2 ($GtCO_2$) (Olivier, Janssens-Maenhout, Muntean, & Peters, 2015), and the CO_2 emission only by Malaysia's electricity generation in 2014 contributed to 0.25% of the Global CO_2 emission (MEIH, 2016). The above analysis further signifies the importance of this research study's contribution to reducing the use of electrical energy, and thus CO_2, by studying the perception of energy experts on Commercial and Industrial Electricity Consumers (CIEC) of the Klang Valley toward the adoption of EE technologies.

However, people are not aware of the consequences and issues regarding environmental problems. This research, therefore, will make significant contributions by providing meaningful information to the population. There are no previous research studies regarding the study of this behavior specific to the population of the Klang Valley in Malaysia (Park & Lee, 2013). Additionally, there has been no research conducted with respect to how energy experts perceive the use of EE technologies by electrical consumers.

Understanding the perspective of energy expert's views can be examined from their experiences in dealing with consumers of commercial and industrial facilities with respect to energy efficiency. This research will influence the subject matter experts in this field to provide guidance on how to get approximately 41,600 commercial and industrial electricity consumers in the Klang Valley (NPIC, 2017) to adopt a stronger EE stance.

In addition, this research will create promising research, development, demonstration, and deployment (RDD&D) opportunities across energy technologies to effectively address the nation's energy needs. More specifically, this research identifies the important RDD&D opportunities across energy supply and end-use in working toward a clean energy economy. The insight gained from this research will provide essential information for decision makers as they develop funding decisions, approaches to public-private partnership, and other strategic actions (DOE, 2015a).

RESEARCH QUESTIONS

This study will address the following research questions:

RQ #1. Why are electricity consumers in the Klang Valley resistant to adopting Energy-Efficient (EE) postures?

RQ #2. How can energy-efficient technologies be implemented in the Klang Valley in Malaysia?

RQ #3. How do consumers find energy-efficient technologies useful?

ASSUMPTIONS AND LIMITATIONS

Most research utilizes delimitations as the boundaries set forth by the researcher to narrow the scope and provide parameters for a study (Creswell, 2014); this study will do the same. This research seeks to explore the issue under study in the specific context of a specific area. This study is limited to exploring the perspectives of energy experts of CIEC in Klang Valley toward the adoption of EE technologies.

The consumers are from all types of organizations, including commercial and industrial sectors, whose final energy demand for fuel is electricity. There are many challenges in implementing EE technologies, and in-depth studies are required that include differences in climate zones, building types, and fuel types, all which play a major role in the types of EE technology needed (DOE, 2015b).

This research is being conducted in Malaysia where there is a tropical climate. However, further research should be conducted in different climate zones and building envelopes as differing climates may require alternative energy solutions. For example, the efficiency of Heating, Ventilation, and Air-Conditioning (HVAC) equipment will differ depending on the climate, type of building, and their thermal conductivity properties when utilizing the same energy efficiency technologies (DOE, 2015b).

Even though equipment can use many other types of fuels, such as liquefied propane gas, kerosene, distillate, geothermal, solar, wood, and natural gas, this research is limited to electricity as the final energy demand fuel. Selection of the fuel is influenced by the feasibility of an EE project, continuous availability, stable unit cost, and the number of equipment that it can efficiently operate by the fuel. Equipment that operates cleanly, efficiently, and safely will dictate the type of fuel selected and if the equipment has the technology to reduce end-use energies. Therefore, a comprehensive study is required based on the final energy demand fuel type.

Furthermore, this study is limited because it does not represent consumers from all types of commercial or industrial facilities utilizing electricity since the performance level of each implemented technology can vary from organization to organization. The capital market contains all kinds of businesses ranging from small to large income companies. In addition, even the work level of a single employee can affect the effectiveness of energy consumption (Carbon Trust, 2018). Therefore, this study assumes that the type of commercial or industrial building and size of the organizations and the product of the organization produces have no bearing on the findings.

This research study is limited to on-demand energies. There are four types of energy usages: end-use energies, useful energies, energy end-use efficiency, and on-demand energies. The energy lost during the supply-demand process is an opportunity missed to reinvest as additional energy resources if the losses could be saved. However, the loss of energy during this process is unavoidable. In addition to increased costs, there will be a corresponding increase in environmental degradation. Raising the energy efficiency at all steps in the supply-demand chain is, of course, the means by which energy losses can be reduced. In the short term, improving energy efficiency directly addresses the so-called avoidable losses, but in the long term, the "unavoidable" losses can be addressed to a degree. For example, a process might be redesigned or an equipment specification may be changed to ensure that the energy losses are kept to a minimum (UNIDO, 2017). Therefore, a further comprehensive study is encouraged to cover the four types of energy use.

Another limitation encountered while undertaking this study was the selection of a small sample of ten highly experienced energy experts from Klang Valley's energy efficiency industry. Energy experts have the same objectives. Their objective is to ensure the Malaysian Regulation of Electricity Supply Act 1990 under the section of Efficient Management of

Electricity Energy Regulation 2008 is compiled mandatorily. For this study, it was assumed that the sample size of ten energy experts would be sufficient to obtain in-depth and meaningful information, which was extracted from energy experts by utilizing open-ended questions during the interview. Hence, this method enabled the researcher to capture the perceptions of energy experts of commercial and industrial electricity consumers and to discover the behavioral factors of electricity consumers that affect adoption and utilization of EE technologies, particularly in the Klang Valley region of Malaysia.

The findings of this phenomenological study are confined to the perceptions and experiences of ten energy experts from the Klang Valley's EE industry. Therefore, the results of this research cannot be generalized for all energy experts in the EE industry. In spite of the limitations mentioned above, this makes a number of noteworthy contributions.

Another limitation this study faced was participant bias, specifically since the data collection was less structured. The participants shared their experiences comfortably and willingly with the researcher and provided data voluntarily; however, it is only human nature not to be able to recall and provide key moments of their vivid experiences that were central to this research. Although it is impossible to eliminate researcher and participant bias, the researcher maintained awareness and vigilance in order to eliminate or reduce such bias, thus ensuring the validity of the findings, especially considering that data collected through qualitative means might have left more room for interpretation compared to the use of numeric data. Furthermore, qualitative research is based on observing and asking questions; hence, it can result in variation in findings. Therefore, different research can produce different findings. As a result, this is one of the major limitations or challenges to the application of this type of research method (Loftus & Higgs, 2010).

Finally, the limited experience of the researcher in carrying out a qualitative phenomenological study was also acknowledged as an additional risk and limitation to this study. As stated by Creswell (2014), limitations are the weaknesses of the study and are beyond the researcher's control. With the guidance of the highly qualified, knowledgeable, and experienced faculty mentor assigned to support this research and the leadership provided by the Swiss Management Centre University Academic Research Board, the impact of possible limitations was reduced.

OPERATIONAL DEFINITIONS

• Industrialization	Industrialization is a process of transformation from human-made work by replacing with machines. For example, turning into an industrial based environment from an agricultural basis is a form of industrialization, or if the necessity to utilize machines becomes high (KeTTHA, 2014).
• Energy consumption	It is defined as the total amount of power used for a specific purpose. For example, if there is a specific energy that is consumed for a use or end-use product, then it is the total energy consumption of a product (The Need Project, 2017).
• Energy-efficient technologies	It is a definition of satisfying energy needed for both demand and supplies. It should be sustainable. In other words, the energy used for a production or a product should not exceed a limit where it extends the usage of power for that particular time or in future (European Commission, 2007).
• Commercial and Industrial Energy Consumers (CIEC)	Commercial and Industrial Energy consumers are energy consumers from the industrial and commercial sectors (Energy Commission of Malaysia, 2016).
• Energy Expert	This a competent person who has met the requirement to be registered as an Electrical Energy Manager as per the Malaysian Regulation of Electricity Supply Act 1990 under the section of Efficient Management of Electricity Energy Regulation 2008 and who is authorized to conduct electrical energy audit and analysis, implementation of efficient energy upgrading program, monitors and keeps records of existing and newly implemented program and to be prompt in submission of information and reports under the regulation (Energy Commission of Malaysia, 2008).

• Sustainable environment	In this context, it can be described as a process of using or saving energy without affecting the present and future generations. The resources in a sustainable environment are created without pollution, resource depletion, and resource harvest and can be used indefinitely without affecting anything. If the resource will not last long, then it should not be continued (RVO, 2017).
• End-user consumption	It is determined as a process of energy consumption where the customer or end-user will use more energy for its consumption (DOE, 2015).
• Energy-efficient practices	Energy-efficient practices are any policies and actions that are taken for energy savings that are considered as energy practices but most of them are merely guidelines and have not been implemented yet (Worrell, et al., 2001).
• On-demand energies	If energy is needed from the end of a production company or industrial end due to the increase in need of the product from the people-end, then it can be termed as on-demand energies (UNIDO, 2017).
• Sustainable development	It is a combination of creating an environment where more policies and rules are implemented from a "top-to-down" approach in implementing sustainable development (Kusek & Rist, 2004).
• Energy labeled products	Mostly, energy labeled products are created by the European Commission and it is used by many companies to market their products if it is energy efficient and make it more valuable (European Commission, 2007).
• Building Envelope	The physical separator between the conditioned and unconditioned environment of a building (Silverman C., 2016).

• Kilotonnes Oil Equivalent (Ktoe)	Tonne of oil equivalent (toe) is a unit of energy defined as the amount of energy released by burning one tonne of crude oil. It is approximately 42 gigajoules or 11,630 kilowatt hours, although as different crude oils have different calorific values, the exact value is defined by convention; several slightly different definitions exist. The toe is sometimes used for large amounts of energy. Multiples of the toe are used, in particular, the ktoes (Ktoe, one thousand toe), megatoe (Mtoe, one million toe), and the gigatoe (Gtoe, one billion toe). A smaller unit of kilogram of oil equivalent (kgoe) is also sometimes used denoting 1/1000 toe (MEIH, 2017).

SUMMARY

This study and its findings are organized in the following manner. The first chapter is an introduction wherein detailed information about the topic is covered. In addition, the research problem is stated, the aim and objectives of the research are outlined, and the scope of the study are explained. The second chapter is the literature review wherein previous research in the context related to the present study are examined and discussed. In this section, a detailed explanation is provided about the significance of implementing EE technologies, a brief explanation about EE technology uses is described, the elements required for research questions are stated, and the objectives that need to be achieved are detailed. The third chapter covers the research methodology used for this study. This section explains the type of research methodology adopted in the present study, which covers the qualitative research methodology utilizing a phenomenology approach. This section discusses the type of research design, data collection, data analysis, and also provides a summary of how the results were supplied. The fourth chapter discusses the results of the study. The results of the present research are covered, and the results derived through methodology are defined. The fifth and final chapter discusses and concludes this study. The chapter discusses the conclusion based on the results of this study which examined to derive what attitudes and behaviors prevent the adaptation of EE technologies. Furthermore, the conclusions of the study are drawn and recommendations for future researches are revealed.

Chapter 2
Literature Review

This chapter provides a literature review containing a critical analysis of the most significant contributions of other authors to the research area and assures that no major information is omitted from previous studies that are related to the adoption of EE technologies. The chapter starts with the explanation of the search for the literature followed by a taxonomy of barriers concluded by other researchers (Armel, 2014; Kadam, 2014; Thollander, Sa, Paramonova, & Cagno, 2015) who discovered through this research study the implementation of EE technologies (IEA, 2011, p. 3), usefulness of EE technologies (UNIDO, 2017), and, lastly, the general behavioral theoretical models associated with the adoption of EE technologies (OECD, 2017). The literature review addresses the three research questions, which provides a clear view of the topic and the elements needed for the adoption of EE technologies These are collectively analyzed based on main and subcategories of barrier groups to determine what is needed for implementing EE technologies and to ensure that all factors or elements that are required to be considered are not neglected before carrying out this research.

RESEARCH PARADIGM ASSUMPTIONS

The section critically discusses the various paradigms that make different assumptions about the social world and identify the one that best fits the present research problem. These differences have been studied extensively (Alvesson & Skoldberg, 2017; Dudovskiy, 2015; Kingdon, 2014), and the following are the four different paradigms most relevant for this research.

The Ontological Assumptions. Ontology is defined as the study of being (Crotty, 2003). It is concerned with "what of the world we are investigating, with the nature of existence, with the structure of reality as such." Guba and Lincoln (1989), stated that the ontological assumptions are those that respond to the question "what is there that can be known?" or "what is the nature of reality?" (p. 83). After understanding the definition of ontology, it now makes sense to identify the ontology for this study where a social world of meanings exists. In this universe, researchers have to assume that the world they investigate is a world populated by human beings who have their own thoughts, interpretations, and meanings. This field of research is clearly manifested in the use of the different research techniques of the interpretive design such as interviews and reaction papers written in a pronunciation course in response to challenging readings in order to interpret feelings and inner thoughts. A realistic ontology is adopted which follows the physical world where the researcher assumes the existence of a world of cause and effect. It is not the ontology of mechanical causes caught in the cause-effect relationships. Pring (2004) referred to this notion saying "one purpose of research is to explain what is the case or what has happened. A reason for seeking explanations might be to predict what will happen in the future or what would happen if there were to be certain interventions" (p. 62). This clearly displays the invitation of research participants (energy experts) to participate in this research study in order to understand their perception of CIEC in Klang Valley toward the adoption of EE technologies.

The Epistemological Assumptions. Epistemology is a method of understanding and explaining how we know and what we know (Silverman, 2015). It is concerned with providing a philosophical grounding for deciding what kinds of knowledge are possible and how we can ensure that they are both adequate and legitimate (Levy, 2006). The epistemological stance used in the first study is constructionism in an effective EE implementation. Constructionism, defined by Crotty (2003), is "the view that all knowledge, and therefore all meaningful reality as such, is contingent upon human practices, being constructed in and out of interaction between human beings and their world, and developed and transmitted within an essentially social context" (p. 42). Thus, meaning is not discovered but constructed. The reason construction is the epistemological stance is because it evaluates the knowledge and awareness of the research participants (energy expert) on the CIEC of Klang Valley posture toward the adoption of EE technology. This then helps in formulating measures to reduce the rate of depletion of natural sources such as fossil fuel and to preserve the environment by reducing GHG and CO_2 emission for their well-being and the well-being of the general population.

The Axiological Assumptions. Axiology in qualitative research addresses value and makes an assumption on how the values of the researcher can influence what is to be studied (value-laden). It also embraces the fact that research is influenced in one way or other by the researcher's value, but at the same time depicts the level of consistency, reliability, or reconstructing of the theories (Alvesson & Skoldberg, 2017). In the context of this study, the participation of energy experts added valuable knowledge regarding the barriers associated with CIEC of Klang Valley in adopting EE technologies. In addition, this research study helps to further improve current policies and regulations in energy reduction, which can mitigate the reduction in GHG and CO_2 emission and protect the environment.

The Rhetorical Assumptions. The art of writing is rhetoric and is referred to as the method of language being employed. It further comprises the art of persuasion and decoration in literature (Sparkes, 2012). Generally, it is deliberated as something to be avoided in research where facts are expressed hypothetically as opposed to speaking for themselves. Scientific writing is a functional process with an equable style that evades enhancement and frequently declares conclusions as either formulas or as propositions. Data presentations of forms are hypothetically substitutable. However, when used in tables, they are contrasted to charts and are considered inconsequential. A standardization form is present, and the format is theory-methods-findings-conclusion, which is being intended in limiting rhetorical excess (Eisner, 1981). Lack in the style reverts to being a rhetorical device in its own right. For example, using proportions is a method for the meaningless language of emotions and influences the reader in the disentanglement of the writer from the analysis. In case one among the threats to the strength of a conclusion arrived from a writer's self-biases, while considering the case of science, any method being projected as a lack of emotion has substantial persuasive power; therefore, language serves as a persuasive function (Marshall & Rossman, 2014). The present research understands the rhetoric of research through the literary criticism of specific reports. Most analyses focus on the language of research and treat data as relatively neutral. Yet the means of data collection, the results of those efforts, and the conventions about how to treat them can be combined to creating specific strategies for persuasion and to project particular images of the research subject. Thus, language does serve a persuasive function in this research. Perception of the research participants (energy experts) on CIEC of the Klang Valley toward the adoption of EE technologies will not be influenced by the researcher when conducting face-to-face interviews to collect the primary data via digital recording. Moreover, they will be transcribed verbatim, and the transcripts will be reviewed by the research participants to ensure accuracy.

The Methodological Assumptions. Methodology is "the strategy, plan of action, process, or design lying behind the choice and use of particular methods and linking the choice and use of the methods to the desired outcomes" (Crotty, 2003, p. 3). Wellington explains methodology as aiming to describe, evaluate, and justify the use of particular methods (as cited in Haig, 2018). The methodology of the first study is case study. Case study is defined by Adelman, Jenkins, and Kemmis (as cited in Crossman, 2017) as the study of an instance in action. It is also defined by Pring (as cited in Linguistics Association, 2017) as "the study of the unique case or the particular instant" (p. 40). Cohen, Manion, and Morrison (as cited in Linguistics Association, 2017) state that the case study methodology provides a unique example of real people in real situations, enabling readers to understand ideas more clearly than simply presenting them with abstract theories (p. 181). The case study used in this study, as explained by Yin (1984), is evaluative (explaining and judging). This is clear in the explanatory case studies in the cross-section analysis. The methodology employed in the second study was the experimental research methodology. This research examines the perception of energy experts on the CIEC of Klang Valley toward the adoption of EE technologies where the primary source of data was from face-to-face interviews and the secondary source of data was from literature review and collection of governmental energy reports. They were transcribed verbatim and analyzed thematically.

THEORETICAL ORIENTATION

The following are reviews on previous researches carried out to analyze and identify the main theoretical framework of this study. The main theories to be discussed are a taxonomy of barriers to energy efficiency, Armel's research barriers, Kadam's research barriers, and Thollander et al.'s research barriers.

Taxonomy of Barriers to Energy Efficiency. Barriers are obstacles that can be overcome with concerted effort, creative management, change in thinking, prioritization, and related shifts in resources, land uses, institutions, etc. (Adger et al., 2009; Ekstrom, Moser, & Torn, 2011). Adger et al. (2009) convincingly argue that many seemly social limits are in fact malleable barriers; they can be overcome with sufficient political will, social support, resources, and effort. However, excessive barriers will make adaptation less efficient, less effective, or require costly changes that lead to missed opportunities or higher costs. In many instances, the barrier may appear to individuals participating in the adaptation process as de facto limits (e.g. a law). Not questioning the changeability of such barriers (however difficult to overcome) may itself be an obstacle to progressing in the adaptation process (Ekstrom, Moser & Torn, 2011). Identification of barriers in EE implementation is vital as it helps to determine the feasibility of an EE project in terms of cost-effectiveness of investment if feasible, and whether the known barrier could be removed, minimized, or overcome (IEA, 2011). After carrying out a comprehensive analysis of peer review literature on previous research associated with the implementation of EE technologies, this section develops a systematic classification of barriers. The following were the main and sub-categories of barriers that were identified in some of the previous research carried out (IEA, 2011).

Armel's Research Barriers. Research conducted by Armel (2014) on behavior and energy states that the main categories of barriers identified were policy, physical environment, sociocultural, interpersonal, and individual. The sub-category barriers for policy were interventions by government and organizations. The sub-category barriers for physical environment were built environment, building, and technology. The sub-category barriers for sociocultural adaptation were communication, habits and myths, implicit marketing, and entertainment education. The sub-category barriers for interpersonal acceptance were face-to-face contact and lack of interpersonal approach. Lastly, the sub-category barriers for individuals were goals, addressing barriers, feedback, and skills. See Table 1 for a summary of the barriers mentioned above.

Table 1 Barriers identified in Armel's research

Main Category Barriers				
Polices	Physical Environment	Sociology	Interpersonal	Individual

Sub-Category Barriers				
Intervention by government	Built environment Building	Communication Habits & Myths	Face-to-Face contact	Goals
Intervention by organizations			Lack of interpersonal approach	Addressing barriers
	Technology	Implicit Marketing		Feedback
		Entertainment Education		Skills

Note: Table derived by author from analysis of Armel's (2014) research.

Kadam's Research Barriers. Kadam (2014) in his research on EE technologies in selected automobile industries identified the main categories of barriers to EE as economical, organizational, and behavioral. The sub-category barriers of the main category economical were management concerns about investment costs of EE measures, difficulty in obtaining financing for EE projects, and limited financial incentive by the government for EE. The sub-category for organizational barriers included concerns that production is given more importance, lack of certified energy auditor/manager in the organization, lack of energy management policies, and lack of energy audits at periodic intervals. Lastly, the sub-category barrier for the main category of behavior was the lack of awareness about energy conservation among the employees and inadequate training programs on energy management. See Table 2 for a summary of the barriers mentioned above.

Table 2 Barriers identified in Kadam's research

Main Categories Barriers		
Economical	Organizational	Behavioral
Sub-Category Barriers		
Management concern Investment cost of EE measure	Core business given priority than EE measures importance	Lack of awareness on energy conservation among the employees
Difficulty in obtaining financing for EE Measures	Lack of certified energy auditor/manager in the organization	Insufficient training programs on energy management

Sub-Category Barriers	
Limited financial incentive by the government for energy efficiency	Lack of policies on energy management
	Lack of periodic energy audits

Note: Table derived by author from analysis of Kadam's (2014) research.

Thollander et al.'s Research Barriers. In a research case study by Thollander, Sa, Paramonova, and Cagno (2015), on EE in the foundry industries in selected European countries, the main categories of barriers were found to be only financial and organizational. The sub-categories for financial barriers were lack of cost reduction exercises, lack of EE investments, lack of employment of performance-based energy services. The sub-category for organizational barriers consisted of a lack of commitment from top management to EE, lack of energy management, lack of long-term energy strategies, lack of future energy policies, lack of energy audits, and a lack of people with real ambitions. Table 3 below summarizes the barriers mentioned above.

Table 3 Barriers identified in Thollander et al.'s research

Main Category Barriers	
Financial	Organizational
Sub-Category Barriers	
Lack of cost reduction exercises	Lack of commitment from top management to energy efficiency
Lack of EE investments	Lack of energy management
Lack of engaging performance-based energy services	Lack of long-term energy strategies
	Lack of future energy policies
	Lack of energy audits
	Lack of people with real ambitions

Note: Table derived by author from analysis of Thollander et al. (2015) research.

This Study's Research Barriers. The study of barriers in EE comprises a multi-disciplinary field with contributions from theoretical backgrounds such as neo-classical economics, organizational economics, behavioral theory, and organizational theory (Sussex, 2017). Theoretically, there are more barriers found in implementing EE technologies that are related to technical, financial, policy, and regulatory which can be addressed if governments take adequate action regarding these issues (UNIDO, 2017). Previous research has identified that adaptability is one of the major barriers to energy efficiency. It was found that environmental, technological, policy, economical, regulatory, and investments were the other barriers (Backman, 2017; Chpassociation, 2015). Based on these theories and the analysis of categorization of barriers from previous research, such as the ones mentioned above, the EE barriers in this study are broadly classified under three main categories: policy and regulatory, economic, financial and marketing, and behavioral, informational and technical.

Table 4 The main and sub-categorization of barriers identified based research reviewed

Main Category Barriers		
Policy and Regulatory	Economic, Finance, and Marketing	Behavioral, Informational, and Technical
Sub-Category Barriers		
No cooperation between agency and organizations (UNIDO, 2017)	Lack of EE gap reduction (Fleiter, Schleich, & Ravivanpong, 2012; World Energy Council, 2013)	Lack of training in energy management (Kadam, 2014; Thollander et al., 2015)
Lack of performance indicators on energy consumption (Armel, 2014; Hiller, Mills, & Reyna, 2012; Thollander et al., 2015)	Imbalance between demand and supply for EE technologies (DOE, 2015a)	Lack of knowledge in EE technologies (Backman, 2017; IEA, 2010; Srinivas, Gadde, Seth, & Dhage, 2015; UNIDO, 2017; & UNEP, 2015)

Sub-Category Barriers		
Strict permitting on environment and technical (Armel, 2014; & DOE, 2015a)	Lack of financial analysis skills in EE (Kadam, 2014; Thollander et al., 2015)	Lack of installation personnel in EE equipment (UNIDO, 2017)
Failure in recognizing regulatory evaluation on EE measures (BEE, 2015; & DOE, 2015a)	Implementation restrictions due to rapid payback requirements (Kadam, 2014; & Bache, 2014)	Lack of standardization in organization's energy management policies (Kadam, 2014; Thollander et al., 2015; UNIDO, 2017)
Control imposed by electricity market (Armel, 2014; DOE, 2015a)	Financial missed opportunities in EE measures (Auffhammer, Blumstein, & Fowlie, 2008; DOE, 2015a; ICC India, 2014)	Lack of awareness in EE among top management (Apeaning & Thollander, 2013)
Lack of government enforcement (APEC, 2011; BEE, 2017; UNEP, 2015; BCA, 2011)	Lack of EE financial recognition (DOE, 2015)	Lack of awareness in EE among employees (Kadam, 2014)
Lack of government incentives (Kadam, 2014)	Disincentives due to lower energy tariff (DOE, 2015a; Energy Commission of Malaysia, 2016)	Energy Management System (EnMS) not integrated into the overall management system of an organization (Kadam, 2014; Thollander et al., 2015; UNIDO, 2017)
Lack of adaptability of government regulations and policies (CCES, 2010; UNIDO, 2017)	Lack of finances for EE training (Kadam, 2014)	Lack of information on organization's own energy trending on consumption (Backman, 2017; Carbon Trust, 2012; DOE, 2015a)

Sub-Category Barriers		
Lack of EE infrastructure (Alcorta, Bazilian, De Simone, & Pedersen, 2014; BEE, 2017; & Thomas, 2012)	Affordability of in-house energy expert (Kadam, 2014)	Lack of information on available EE technologies (Armel, 2014; Auffhammer, Blumstein, & Fowlie, 2008; European Commission, 2017; & Murphy & Harris, 2014)
Industrial regulation on utility usage (Armel, 2014; DOE, 2015a)	Low priority given for EE projects (Apeaning & Thollander, 2013; ICC India, 2014; Kadam, 2014)	
Lack of policies on energy management (CCES, 2010; Kadam, 2014)	High financial risk of implementing EE projects (DOE, 2015a; & Sussex, 2017)	
Weak organizational policies on energy efficiency (Kadam, 2014)	Lack of internal and external capital for EE projects (Apeaning R., 2012; Thollander et al., 2015)	
Lack of policies on secondary effects of GHG or CO_2 emission (DOE, 2015a)	Organizations limitations on EE due to budget policies (Thollander et al., 2015)	
	Lack of cost reduction in organization's policies (Thollander et al., 2015)	
	Lack of measuring system for key EE performance indicator (Backman, 2017; Thollander et al., 2015)	

Sub-Category	Barriers	
	Lack of EE adaptation even with attractive financial returns (Backman, 2017; DOE, 2015a)	
	Lack of financial incentives from the government (Kadam, 2014)	
	Perceived Marginal Benefits (Armel, 2014; EPA, 2017)	

Note: Table derived by author from analysis of literature review.

General Barrier in Implementing Energy Efficiency Technologies. The potential to increase EE are tremendous, but they are usually ignored since the opportunities to implement EE solution are deterred by some critical factors. These critical factored are called barriers comprised of all factors that create obstacles in the adoption of cost-effective energy-efficient technologies or slows market penetration (Fleiter, Schleich, & Ravivanpong, 2012).

EE Technology – Barrier. According to Cagno and Trianni (2013), barriers in implementing EE technologies have been identified as far back as the 1970s, and it is not new in literatures related to this subject. The first attempt in classifying barriers into six categories was made by Auffhammer, Blumstein, and Fowlie (2008). They identified the following six categories: lack of information, market structure, custom, misplaced incentives, regulation, and financing. Montalvo (2008) was one of the few researchers who had classified the barriers in an attempt to develop and utilize a cleaner environment. He mentioned the beneficial factors. However, there are both positive and negative outcomes, which depend on time, context, and circumstances. If a consumer is ready to implement EE technology, a clear distinction between both external and internal factors has to be made. This can help organizations efficiently design and implement eco-friendly products once the barriers are identified (Cagno, Worrell, Trianni, &

Pugliese, 2013). If energy wastage is properly managed, then incorporating government policies can solve most of the issues. Chan, Qian, and Lam (2009) affirm that government objectives and commercial architects' viewpoints can be linked.

The viewpoints of building designers influence studies being conducted based on creating a green economy. The benefits which can be obtained in implementing sustainable green building practices is directly dependent on the role of shareholders which can strengthen competitors' advantage and protect the reputation of enterprises. This could also turn out to be an advantageous action since it could lead to the development of routines and a set of beliefs that can be assembled into operations for a carbon reduction strategy (Tan, Shen, & Yao, 2011; Wong, Ng, & Shahidi, 2013). There has been some research carried out on enterprises of different sizes. A survey questionnaire by Akadiri and Fadiya (2013) was conducted to target green practitioners, architects, construction managers, structural engineers, and quantity surveyors, especially in the construction industry. The study determined that there is a significant relationship between the adaptation of sustainable construction practices and the firm's size; however, the previous research considered a large sample size to be associated with large-scale firms, and hence, the findings of the research could be generalized more for large-scale organizations. Zhu and Geng (2013), stated from a different viewpoint that a firm's size does not create barriers in extending supply chain practices related to energy saving and reducing carbon emission.

Another survey classified the respondents based on their designation in order to analyze the various viewpoints based on their position in the organization. The survey satisfied the criticism gap that only one person was chosen from an organization and it ignored the impact of other designations and their influence on barriers. The studies done by Davies and Osmani (2011), Kostka, Moslener, and Andreas (2013), Osmani and O'Reilly (2009), and Wang, Li, and Tam (2014) analyzed different perceptions. Managers of contracting organizations were chosen since they have more experience in projects pertaining to constraints, sustainability, and environmental management. In an empirical study by Wang, Li, and Tam (2014), the employees chosen for the study must have at least worked in the design of a construction-based project in the building sector. A researcher is then able to identify the number of employees chosen who were directly involved in the number of sustainable research projects. The study has deduced that employees are aware of the steps taken for implementing sustainable development. Although there have been various studies conducted in this field of research, further studies still need to be conducted. At this moment, it can only be concluded that the research method utilized was based on subjective judgment and experience (Kusek & Rist, 2004).

Determination of Barrier. Barriers can be determined only if they are justified based on proof of the market failure of energy efficiency technologies in real-world application. There is, though, a possibility that not all barriers can be identified. This occurs when there is a lack of implementation of EE technologies due to there being a minimal necessity for intervention due to the likelihood of minimal market failures. Some of the common barriers currently known in the industry are operation cost of capital markets and hidden costs (Sussex, 2017). The expenditure from the government, regulatory burdens, and level of utility have some effect on a product's efficiency performance. There are a wide range and multiple types of EE products. Some EE products are purchased without having the correct information on its technical performance. Products such as lights, insulation, boilers, and coolers have no frequency of purchase information (e.g., life span estimate) (OECD, 1996). Even if EE technologies have been implemented, customers will have difficulty in evaluating the performance of a system such as motors, speed drivers, and control systems. This lack of actual performance data regarding the purchases has an effect on customer's behavior. Utilization of energy should be calculated from a low-level sub-metering which is supposed to be available at every building. The sub-metering will be beneficial since it can give the cost of monthly energy before and after implementing EE technologies, and the difference between these costs will then be the energy savings (Carbon Trust, 2012).

A report by DOE (2015a) categorized industrial EE technologies and measures into three types: end-use, demand response, and combined heat and power. These pose barriers because it is unpredictable to analyze the benefits or efficiency, as there are no known standard methods or procedures to determine the benefits or efficiency in each category. For example, if an organization is not able to establish the EE of a technology, then the possibility of it being implemented is minimal. Economically and financially, there are many barriers preventing consumers from implementing EE technologies, such as tax structures, split incentives, lack of economic demand, internal competition within capitals, time-based rates, accounting practices, sales outside the capitals, failure in recognition, and affordable energy prices. Some barriers may be resolved by regulatory policies when there is a failure in implementing EE practices, lack of institutional participation, lack of involvement from employee, and failure in carrying out environmental protection. Lack of standardization, cost recovery, market demand, and exclusion from the resource standard are considered barriers for an organization when implementing EE technologies. The last exclusion from the resource standard comes with informational barriers consisting of a lack of awareness, lack of data on energy consumption, lack of technical knowledge, and administrative burden. The barriers mentioned are economic, regulatory, and informational barriers. These barriers affect the implementation of technologies such as gasification combined cycle coal technology, pulverized fluidized bed combustion, and biomass integrated gasification combined cycle (DOE, 2015a). Each technology has been revised in a survey to identify the involvement of major stakeholders related to the Indian power sectors utilizing an analytic hierarchy process in finding the level of adaptability (PWC, 2016). Both developed and developing countries take more time since the speed of differentiating the type of business is vast and it can only be processed slowly with time. The implementation of government policy and the support from the government will play a major part in evaluating the value of EE measures. If the price of energy is still high after implementing EE technologies, then the organization will plan to generate its own energy, aiming for long-term benefits. The government and its agencies must directly disseminate and share with the public information related to EE. The machinery and appliances should be labeled or certified, indicating that the equipment is EE rated as a standard practice. The Indian Bureau of Energy Efficiency claims that a certified product with EE rating carries more value when implementing EE projects (BEE, 2017).

Electricity production and its cost differ for each country depending on energy demand and its affordability by population. Researchers have calculated the average value of electricity when deriving the whole results; this can act as a significant barrier. When interpreting with scientific-based potentials, every aspect in which the technologies are applied should be considered. Each plant might have its own factors in determining the cost of electricity. Moreover, surveys to analyze these factors affecting the cost of electricity have not been carried out until now (UNICEF, 2017). The results from these studies are analyzed using lifetime data. It is quite difficult to determine the future cost of electricity as the life-cycle cost approach in implementing EE technologies takes into account the energy saved discounted over the life of the technology employing the Net Present Value (NPV) analysis. This poses a barrier in determining the actual success of the EE project. The sum of investment costs and the operating costs of the EE technology discounted over the lifetime (NPV) is taken to derive the return on investment (ROI) of the EE project. In general, EE projects aim at achieving the ROI within a three-year requirement, for example, the payback period and to fulfill an acceptable Internal Rate of Return (IRR) for the investment. The life cycle can comprise of a top-down or bottom-up approach. This type of organization can use either one of the two based on the availability of data to select the best approach and be diligent on how the lifetime data analyzed is utilized. Realizable and realistic market potential cannot be termed the same. A realizable market potential is based on price, energy potential, and energy savings. Realistic market potential comprises non-financial barriers such as acceptance of solutions and business cultures (World Energy Council, 2013).

Policy and Regulatory Barriers. Barriers have been found in implementing EE technologies that are related to policy and regulations. Regulatory and policy-based systems are actually influenced by certain manners or priorities that will influence EE measures. In the case of policies, both national and local government guidelines are included. In many countries, especially in Africa, there is no policy. If any policy is in place, it can be indifferent and counter-productive to energy efficiency. Africa has the lowest levels of access to basic energy supplies. Even moderate energy consumption can often have very high development benefits. By using energy more efficiently, African nations can maximize the effective use of available resources for the economic benefit of their populations. According to the finding of UNIDO (2017), there have been various barriers in regard to this, such as insufficient cooperation among researchers or research organizations, making it difficult to implement an effective EE research and in implementing development and demonstration programs. Even if the frameworks provided are effective, transferring research prototypes of EE products into actual working products proves to be challenging.

According to Hiller, Miller, and Reyna (2012), if the energy performance metrics are not clearly defined, then there is no possibility of finding out the actual outcome of implementing EE projects. Without knowing this outcome, it becomes difficult to gain support within and outside of an organization to implement further EE projects, thus creating a barrier.

In the study carried out by DOE (2015a), environmental and regulatory issues are mostly concerned with emission, and if it is inclined toward a new source, then a review to permit its usage is required. The implementation of combined heat and power (CHP) (see Figure 6) equipment can reduce potential carbon emission based on regulatory evaluation. The failure to recognize the regulatory evaluations may cause many companies procurement to omit resources in their utility plan, thus reducing the revenue stream. Standby rates are defined as a structure of rates derived from the project's capital investment and operating cost of the CHP. In the United States of America, a Clean Energy Portfolio Standard (CEPS) program has exclusion standards with respect to clean energy. CHP's eligibility is actually under the representation of CEPS programs consisting of combined standards from Renewable Portfolio Standards (RPS), Energy Efficiency Resource Standard (EERS), and Alternative Energy Portfolio Standard (AEPS). The growth of CHP is limited by the combined standards, which impose strict environmental and technical requirements. Additionally, complication brought about by limits in electricity markets creates barriers (DOE, 2015b).

There is difficulty in adapting the solutions that are given to private organizations for EE measures because they require a lot of change. The solutions need to have a balance between the external measures, such as governmental regulations and policies, and internal measures, such as the private organization's goals and objectives. External measures are depended on the collaboration of one or more themes from governmental and non-governmental entities. State and federal governments have their guidelines based on various policies. Public policies are the ones that can be applied by federal and state agencies in offering grants, loans, regulatory programs, and other financial incentives (CCES, 2010).

An inappropriate energy tariff supported by regulations can limit the interest of energy efficiency. It is common for large energy consumers to receive discounts on their energy tariff, which acts as a disincentive to implement EE measures (UNIDO, 2017). However, removal of energy subsidies could be the ample recommendation to increase the use of energy-efficient concepts and reduce the over-consumption of natural fuel resources such as coal, natural gas, and fossil fuel (Energy Commission of Malaysia, 2016).

Lack of implementation of policies of EE programs was discovered. This was due to governmental agencies and their local Department of Energy (DOE). The objective of implementing an energy management program is to reduce the government's monthly energy consumption, mainly on government-operated vehicles and for various other measures such as efficient lighting, fuel-efficient switching, and air-conditioning retrofitting of their buildings. The Department of Science and Technology (DOST) collaborated with many private and government organizations in labeling energy rating on commercial and industrial equipment. Established MEPS were implemented on air-conditioning and refrigerators and, in the near future, on television. DOE also includes many other programs that are being implemented through its various agencies. The government's EE drive is evident in the drawing up of the Energy Efficiency and Conservation Roadmap for years 2012 to 2030. This led to the enactment of the EE and Conservation Act in 2014 (UNEP, 2015).

In other countries, different types of barriers to energy efficiency can be found. In India, for example, in 2007 the government had developed an Energy Conservation Building Code (ECBC) for new commercial building standards to be implemented on a voluntary basis until made mandatory by individual state governments. In 2008, the code was updated to include buildings with a connected load of 100kW and above as compared to 500kW and above requirement in the 2007 version. As of 2012, ten of the 28 states in India have made the ECBC mandatory, and six others have initiated the amendment and notification process for mandating the code (BEE, 2017).

There are three main barriers to code implementation. The first is the technical barriers. The market for the EE products (glazing, chillers, and insulation) is still not developed, and product availability to meet the code requirements is a major barrier. The associated barrier is inadequate testing facilities to certify products as per code requirement. The technical capabilities of code implementing agencies are not adequate to support code implementation and verification. There are other codes in the South East Asian (SEA) nations such as the EE Building Guidelines and EE and Conservation Initiative Awards Scheme in Brunei Darussalam; Promotion of EE and Conservation (PROMEEC) in Cambodia; GREENSHIP Building

Rating Tool, Energy Benchmark, and Best Practice Guide in Indonesia; and Energy Rating and Labelling Programme, Malaysian Standard Code of Practice on the Use of Renewable Energy, EE in Non-Residential Buildings (Pheng, 2007), Green Building Index (GBI), and Energy Performance Management Systems (EPMS) energy audits on government buildings in Malaysia (APEC, 2011; UNEP & BCA, 2011). However, these codes are not mandatory and hence knowledge toward these schemes among energy consumers is lacking. There is a lack of knowledge among designers to analyze designs based on code requirements. Energy simulation capability to quantify savings based on EE parameters as defined by the code is very limited, which is revealed in the reports by Srinivas, Gadde, Seth, and Dhage (2015) and KeTTHA (2014). The building construction industry (contractors and services providers) is not competent to apply these measures on-site. The energy conservation act empowers each state government to amend the ECB Codes to suit each state's particular regional and local climatic conditions. This provision may, in the long run, lead to large deviations from the ECBC that has been developed by the Bureau of Energy Efficiency (BEE). This may lead to confusion among builders, developers, and designers. The second obstacle is institutional barriers. Although there are state-designated agencies for code implementation, the institutional framework to support the same is yet to be in place. There is no defined mechanism within the states as to how the code should be enforced. The third difficult barrier is financial. To boost the promotion of EE services, the government will need to relax import duty and reduce tax and excise duty. The government could play a major role in eliminating these financial barriers (Majumdar, 2010).

Moreover, if there is no independent power producer, such as CHPs, then there is no possibility to sell electricity to the grid, which could improve efficiency within the electricity grid system. Likewise, the lack of access to a gas network would not allow using gas-based EE technologies. Reduction of GHG will require oil and coal to be substituted with gas, which requires access to gas grids. Technically, implementation of energy infrastructures is characterized by decreasing marginal costs that are usually associated with natural monopoly. In this case, access to technical network and pricing may have to be regulated. Lack of technical infrastructure may be the result of regulatory failure. For example, regulators need to ensure that investors or operators will be able to recover costs, or they will not invest or carry out operation (Alcorta, Bazilian, De Simone & Pedersen, 2014).

Industrial regulations have a direct impact on a country's utility regulation and how utility is used. Due to a lack of industrial regulations, consumers will not have a positive attitude toward EE implementation, which will provide savings on their utility expenditure. Consumers can expect higher savings than usual from EE projects as designers of these projects will normally use the minimum savings as the performance guarantee, and a successful energy savings project will have an effect on the consumer to consider further investment on EE in order to decrease their energy supply requirements.

The design of policy relating to incentive must take into careful consideration the interests of the private and public sectors since the policy covers both public (reduction of greenhouse gases) and private (energy cost savings) benefits. Governmental energy management programs are crucial in overcoming the general barriers, typically the lack of awareness on the implementation of an energy management system (EnMS). Governments must provide support and guidance throughout the implementation process (Kelley, Goldberg, Magdon-Ismail, Mertsalov & Wallace, 2011).

In general, the range of equipment and types of energy-efficient technologies in the domestic markets are low. Equipment and technologies necessary for EE projects need to be imported, and there are many barriers in applying EE measures in industries and businesses. Considering the barriers faced by different nations, it is found that there is a lack of knowledge and capacity in incorporating EE measures among local governments, such as in reducing or managing the losses in transmission and distribution of electrical power to consumers (IEA, 2010). From the analysis of data by Kadam (2014), due to lack of awareness and training in energy management, manufacturing organizations are unable to formulate strong EE policies. Due to this, in a manufacturing environment, production is given more importance than energy management.

Lack of a Legislative Framework. Climate proofing is required due to global warming, which is mainly caused by GHG emission from the combustion of fossil fuels in cars, factories, and electricity production. As global warming progresses, the temperature of the planet is expected to rise by around 4°C by the year 2100. This increase is not evenly distributed around the globe, with the polar ice caps experiencing the biggest temperature rise from the measurement received from localized feedback mechanisms. This increase in temperature is already leading to the polar ice caps melting at a faster rate than predicted, and this will ultimately lead to a rise in sea level, leaving some low-lying countries in real danger of being destroyed. However, it is not only sea level that will be affected by the warmer planet, but also many other associated changes to precipitation will occur, leading to flooding in some areas and droughts in others. To combat these changes, architects, town planners, and engineers are starting to focus their attention on 'climate proofing', which is the practice of making buildings and infrastructure usable if changes to the environment continue (Thomas, 2012). Climate proofing requires the implementation of regulatory practices and applying a definite framework. However, most of the failures are due to poor adaptation and reduced business interest or competitiveness within them. There are insufficient legal frameworks that address gaps in preventing climate change (Murphy & Harris, 2014).

Economic, Financial, and Market Barriers. In many cases, EE technologies that have been implemented utilize technologies that are matured, proven, and economic. The main challenge in implementing policy to reduce the EE gap with proven and economic EE technologies is being able to increase the rate of implementation. The EE gap is the space between what is economical and what EE technologies have been implemented. In maintaining credibility and the drive for investment, the policies implemented should be consistent. Implementing EE measures in an industry is a long-term process, and to achieve it, a long-term target is required. For example, Thailand has set a 20 years target for its EE development plan, which is to be implemented starting from 2011 and ending in 2030 (Fawkes, Oung, & Thorpe, 2016).

From an industrial point of view, even with advancement in EE technology, there will be gaps in managing the implementation of EE technologies in the real world. From an economic point of view, barriers are created when there is a higher supply of EE equipment compared to demand. The lack of demand is due to the low availability of energy consuming systems requiring the adaptation of EE technology. The insufficient amount of energy consuming systems that require the adaptation of EE technologies will create a weak market for EE technology. There will be no incentives for manufacturers and importers to promote their EE products in this market, as it will not be economically feasible for the manufacturer or for the importer to maintain a presence in this market.

This lack of market for EE technology will lead to continuous utilization of traditional and obsolete energy inefficient equipment, thus wasting energy. In Africa, the success of the implementation of EE technology is further hindered by unethical marketing by misrepresenting EE product capabilities, which led to EE project failures (UNIDO, 2017).

If an organization has to implement an EE project, then the implementer should be aware of which project he or she must prioritize, that is, which project provided the highest success in terms of ROI. The report by EDF Climate Corps indicates that up to 31% energy savings can be achieved through the implementation of EE technologies. Organizations investing in EE categorize the type of energy consuming services in their facility such as lighting, compressed air, motors and HVAC, vertical transportation, and so forth. According to the International Chamber Commerce of India (2014), it was reported that 70–80% of the opportunities to achieve large-scale energy efficiency exist in only four technology groups: lighting, compressed air, motors, and heating and cooling (ICC India, 2014). Priority to implement EE project is based on rapid payback, high probability of success, and few negative consequences posing barriers. The reason for this strategy is that the savings from rapid payback projects could finance the lower priority projects, hence lowering the company's total capital investment in EE projects (Hiller, Mills, & Reyna, 2012). Achieving rapid payback is a barrier as the upfront costs are high for EE projects. Initial high capital outlay becomes a barrier for most organizations, especially with Small Medium Enterprises (SMEs), as they may not have ready access to capital (Bache, 2014).

The report showed that a utility-based model could be applied if the cost of capital recovery and a lower cost to produce energy favor the installation of industrial CHP (cogeneration system), which most consumers are unaware of, and as a result, this becomes a missed financial opportunity. Figure 6 illustrates a simple schematic of a CHP system. CHP is an on-site electricity generation that captures the heat that would otherwise be wasted, which could provide useful thermal energy such as steam or hot water that could be used for space heating, cooling, domestic hot water, and for industrial process apart from producing electricity. In this way, and by avoiding distribution losses, CHP can achieve efficiencies of over 80 percent, compared to 50 percent for conventional technologies (e.g. grid-supplied electricity and on-site boiler). There is also a lack of recognition from financial reports on EE technologies' beneficial contribution to the overall environmental (EPA, 2017).

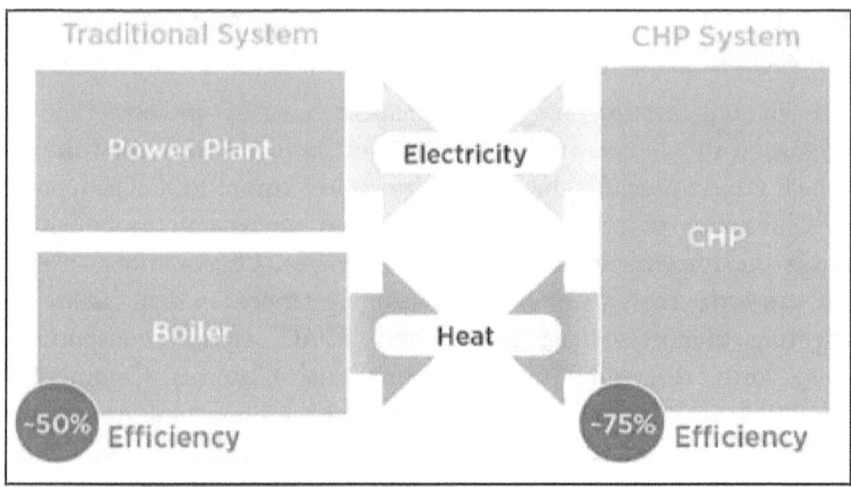

Figure 6 Combined heat and power system: efficiencies. Adapted from "Combined Heat and Power: A Clean Energy Solution," by U.S. Environmental Protection Agency's Combined Heat and Power Partnership (2015).

The effect of a lower tariff price structure can act as a great discouragement in implementing CHP technologies according to Fawkes (2016). The author further iterates that comprehensive EE policies, which combine energy supply regulation and industrial EE requirements, should be able to tackle the issues mentioned earlier (Fawkes, Oung, & Thorpe, 2016).

According to the findings by UNIDO (2017), the availability of skilled and knowledgeable staff in energy management are limited due to affordability in small and medium industries. Skills required to carry out EE projects in these industries could be outsourced to external energy management experts to implement their in-house EE projects in order to overcome the above-mentioned barrier.

The research carried out by Apeaning and Thollander (2013) suggest that the prioritization of investment for EE projects is ranked low among top managers, which creates a barrier. If top management of an organization perceives the implementing of EE technologies as low priority, then the EE project will be treated as a secondary option with respect to investment.

Top management of an organization would rather invest in operational equipment to provide fail-safe measures requiring low investment such as a generator as a source of standby electricity supply in the event of a disruption in electricity supplied by the grid. The rationale behind top management's decision to consider the implementation of EE projects as a low priority is due to uncertainty and risk.

Many energy savings opportunities in manufacturing sectors apply to primary processes having continuous operations according to the study carried out by ICF Consulting (2012). There would be a significant financial loss if production was disrupted to implement EE technology. Even though energy is a significant portion of production cost, a frequent manufacturing plant will optimize the process to maximize production and not energy consumption. The implementation of EE measures is planned around equipment investment cycles rather than short term EE opportunities.

According to Sorrell, Mallett, and Nye (2011), problems associated with capital can be divided into two types of barriers. One is due to the low priority given to EE technology investment, and the second involves budget policies and procedures. The authors further iterate that due to insufficient funds within internal organizations, raising additional shares to generate funds and borrowing can be an issue. Almost all the studies referred to have different business situations. An example of this is when an industry faces declining demand for its product, reduced margins, overcapacity, and increased competition, which could lead to site closures, cost-cutting exercises, and staff reduction. Internal funds within the organization are restricted due to priority issues, and every organization has to undergo proper budget planning to make it effective. It is found that budget policies and procedures of each organization differ in process and decision making, and not all will have the same outcome, especially in prioritizing EE projects within their budgets. Due to this fact, many organizations will always try to delay small projects or EE projects if the budget does not permit. Management's objective is to improve the organization's business and maximize profits for the shareholders by increasing the turnover of the organization where non-discretionary projects requiring investment, such as EE projects, will be prioritized based on essentiality. Management in organizations fail to realize that savings contributed from an EE project will directly improve business profits, but the barrier hindering the management in investing in EE project is the initial capital outlay which does not favor the business in the short term (Sorrell, Mallett, & Nye, 2011). This hence requires management personnel in organizations to empower themselves on improving the profitability of organizations using EE policies which are deemed to generate better results in the long term.

According to Backman (2017), in Sweden, approximately 55% of the companies did not invest in EE projects, which has a 19% return on investment. Limited access to capital is a major barrier with respect to EE investment. However, several companies that have access to capital had successful EE projects. The author further implied that a limitation of investment in EE projects should not be considered as a significant barrier, and this limitation should be the organizations' responsibility to overcome. EE investment does not always conform to a specific budget because each EE measure has its own characteristics. Cost reduction should be the driver in implementing EE projects and should be the key performance indicator to be monitored and controlled continuously (Backman, 2017).

According to Beuc (2015), the public has the best interest in industries or business managing EE investments, and the interest in the implementation of EE is dependent on the willingness of the industry and business owners. The attractive return on investments in EE can make the business owner adapt EE technologies while carrying out aesthetic improvements to their working conditions. The consumers do not have full control over market demand in order to reflect their preferences. Hence, policy cannot be implemented in addressing the cases toward market failure in the above-mentioned circumstance.

Financial barriers play a significant role in limiting energy efficiency investments, mainly due to a lack of financial incentives for energy-efficient technologies according to Srinivas, Gadde, Seth, and Dhage (2015). EE in buildings is not given due consideration in funding and incentive by the government. Proper revision in regulation frameworks needed to be made by providing incentives, duty relaxation, and tax benefits. Investment in EE does not seem to be lucrative to financial institutes due to uncertainty about the returns. Promoting EE in buildings will require innovative financial schemes. Some reasons why motivating financial schemes are not working are lack of awareness in short amortization cycle, lack of incentive for investors and contractors to build ECBC compliant buildings, and/or a lack of awareness that low electricity bills could be a powerful marketing tool for future rental contracts. The high cost of borrowing impedes incremental funding in EE that could be offset with future energy cost savings.

According to Kelley, Goldberg, Magdon-Ismail, Mertsalov, and Wallace (2011), once a company successfully implements an energy management system (EnMS), then the information it provides will in turn address a variety of other common barriers to EE such as the perceived technical risks and financial viability of EE projects. The EE market will have to correlate with finances and the perceived technical risks that will likely occur when implementing EE projects. The experience of the market in accepting an energy management system associated with government-led programs to stimulate and encourage companies to implement them. This is further confirmed by a study by IEA (2013) that if the overall EnMS system was successfully implemented with proper association with both government and private organizations, then the profit that will be gained will be cost-effective. It is a fact that covering only industrial sectors will reduce energy consumption by about 20% if technical-based efficiency is applied. The significance of this is made clear by the EIA (2014) report stating the fact that about 26% of the CO_2 emissions are dependent on the potential of the industrial sectors.

According to Murphy and Harris (2014), there is a lack of technical information and assistance in addressing climate change, and due to this, there is a limitation in carbon financing to help project developers. There is a lack of climate resilience and financial assistance in addressing these issues. Private organizations have low carbon space in terms of financial services. The classification provided by the Kenyan financial institutions is not specific to low carbon investment and is expensive. Banks are the major barriers as they do not understand the many long-term opportunities since they require high levels of collateral, affordable credit, and are reluctant to lend to small and medium-sized organizations to necessitate climate proofing.

In a study conducted, it was shown that the cause of low implementation of cost-effective EE measures in firms principally stems from rational behavior to economic barriers, which are deeply rooted in the lack of government frameworks for industrial energy efficiency. The respondents of the survey identified that limited access to internal and external funds was the most important obstacle preventing EE measures to be implemented. Internal access to funds is limited by the low awareness of top management to EE improvement measures, which in effect results in low priority of EE investments. On the other hand, external access to funds is limited by high interest rates associated with loans from banks and financial institutes. Most of the surveyed industries had no energy management policy (Apeaning, 2012).

EE investments are seen as marginal in the context of unfamiliar complexity and effort required. This situation applies to small and medium-sized companies (SME) where the implementation rates of EE programs are lower than the market expectation. Manufacturing sectors prefer to invest available capital and resources in improving their manufacturing processes rather than in EE implementation because this area of cost savings is outside of their expertise, seen as risky, and requires tremendous effort to implement. Apart from the importance given by government policymakers and environmental lobbyists, EE is regarded as a non-key priority in many sectors. However, in actuality, EE has the largest potential compared to other energy savings measures while reducing carbon emission (ICC India, 2014).

Behavioral, Information, and Technical Barrier. A report by UNIDO (2017) confirms that engaging installation personnel who are inadequately trained in implementing EE technologies will influence the behavior of customers in reducing their lack of confidence in the project's success. This is further substantiated for the analysis of data by Kadam (2014). The data shows that there is a lack of awareness in applying energy management techniques as well as insufficient training programs in energy management. The employees are the major barriers to the adaptation of EE technologies. From a report by EDF Climate Corps (2017), the companies that were surveyed had little or no knowledge on how an EE technology affects their performance in terms of energy use. There is also a lack of information and incomplete standardization related to the implementation of CHP (DOE, 2015).

Apeaning and Thollander (2013) affirm that regardless of many guidelines and policies with respect to energy efficiency, most top management is not aware of their responsibilities toward the implementation of EE projects. The authors further iterated that low awareness of the benefits of EE within the top management of companies is reflected in the behavior of their own employees toward energy savings. Energy management requires staff within an organization to acquire trained skills and knowledge in EE technologies and projects. This is apparently lacking in many small and medium industries.

According to a report by UNIDO (2017), the process of managing energy demands in an organization is called energy management and has many components. One such component is project management, which is a continuous process requiring an association with the total management system of an organization. In managing an EE project, the organization must make energy management an integral part of the organization's total management system in order to address the issues related to economics, finance, maintenance, production, safety, health, and environment. In most cases, the integration between the organization's total management system and energy management is neglected.

In the manufacturing sector, there is a broad range of end-use EE technology that could be implemented to mitigate energy utilization according to DOE (2015b). There are diverse possibilities in applying EE measures, for instance, replacing existing equipment to high efficient ones such as high efficient boilers, electric motors, lamps, and so forth. In order to manage energy usage optimally, there is a need to implement a real-time computer-based control and energy monitoring system. The energy management system, with its real-time data acquisition system and with the appropriate analytical tools, can optimize energy usage of energy consuming equipment in the manufacturing facilities and provides established statistical reports on the performance of the organization's EE initiatives. These statistical reports can provide valuable information to further improve the existing EE projects and to promote new measures (DOE, 2015a). In most cases, the above-mentioned energy management systems are lacking in the manufacturing sectors.

According to CCES (2010), a less expensive way of implementing EE is by reducing GHG emissions, which comes with reduced risk and operational costs. Many companies reduce the GHG emission gap by implementing other companies' effective EE technologies. The barrier is in finding the right skills needed for implementing EE that can help in achieving sustainable energy savings requiring common knowledge and skills pertaining to initiating an energy management program. No single person on the energy team is expected to be proficient in all areas of energy management, but members of a successful team should collectively possess the following knowledge and skills:

- Management skills
  - Business decision-making fundamentals
  - Business improvement skills
  - Organizational and leadership skills
- Knowledge of regulations, standards, and best practices
  - Federal, state, and local energy legislation and policies
  - National energy reporting systems
  - Federal, state, and local green building standards and programs on environmental regulations
  - Energy management system concepts (e.g., ISO 50001)
- Financial and accounting skills
  - Financial decision-making processes
  - Risk management
  - Economics of energy management
- Technical knowledge
  - Facility and industrial processes
  - Energy fundamentals
  - Energy metrics
  - Energy measurement and verification techniques and protocols (Lunt, Ball, & Levers, 2014)
- Other knowledge and skills such as communication and interpersonal skills (GSEP, 2013, p. 9)

There are more unforeseen barriers when initiatives are taken to reduce energy. The barriers could be the information provided on implementing EE technologies to both private and public organizations (CCES, 2010). Another barrier could be that the solutions given may have originated from different types of industries such as services, production, operations, or supply chain sectors that might not be readily adaptable to other industrial sectors.

Backman (2017) stresses that an energy audit does not affect the investment plans directly, and investment plans and energy audits should be treated separately. Lack of technological information and funds are major barriers in overcoming municipal energy policies. Most researchers agree that the main barriers in increasing EE measures are due to the lack of information in various areas. Some of the information lacking are the real cost of implementing new EE technologies, the total cost of the consumer's own energy consumption, knowledge on EE technologies, and the cost of operating and maintaining EE equipment. If these costs are not taken into consideration, the particular EE project will become a high-risk investment if the aim is to have high returns (European Commission, 2007). These non-transparencies of information can be misleading to an organization, especially in conducting the feasibility of investing in an EE project.

A report by the European Commission (2017) mentioned that it should be mandatory to share information pertaining to the implementation of EE technologies by disseminating it to the public. The purchase and utilization of EE products will reduce the impact due to the lack of unbiased information and lack of credibility within an organization. It does not only rest upon private organizations, but the government and other industries should be committed to EE measures. Public awareness is required in order to understand the necessary steps to be taken and different options in the process of taking EE measures. As reported, the government has the greater responsibility to promote EE success stories compared to the private sectors. Every energy consuming product must be labeled to provide information on its EE rating or performance. The energy saved by implementing EE measures will outweigh the additional cost of EE equipment compared to non-EE equipment and, to a certain point, will offset future energy increases. Providing unique identification labeling on EE products will encourage the purchase of EE products.

When the European Commission questioned its committee members on how the general community will have a better understanding of EE technology investment, a committee member replied by stating that most EE technologies based on demonstration results are worthy of investment. In actual fact, after the implementation of the technology on-site, it fails to deliver the expected results (Commission of the European Communities,

2005). According to Schleich and Gruber (2008), the issue lies in the lack of information shared by the consumers on their actual problems faced with the EE technologies in part due to the lack of sharing information on the cost of the technology. The authors further iterated that if a commissioned survey were carried out to evaluate the performance of EE technologies, it would be unclear if the results would show a correlation with the problem discussed above. Out of all barriers deduced by a majority of researchers, information is the main barrier, which was recommended for further research.

In a study carried out by Backman (2017), decision makers in small organizations were found to be ill-equipped and ill-informed in making EE implementation decisions. At the heart of the information barrier is the transaction cost, which is the cost of gathering, assessing, and applying information about energy savings potentials and measures, as well as the cost associated with finding and negotiating the contracts with potential suppliers, consultants, or installers, or the cost of implementing, monitoring, and enforcing contracts. Since EE measures are ideologically and technologically complex, the cost of transaction will be above average, and as a result, the information barrier will be more significant in EE technologies than the other technologies. Hence, to overcome the information barrier in EE technologies, a study conducted by Tremblay, Lalancette, and Roseveare (2012) concluded that the effective supply of relevant information of the right quality and the education and training of the consumer are important contributions in overcoming the barrier posed by the lack of technical capacity. This was further confirmed by Dunstan, Daly, Langham, Boronyak, and Rutovitz (2011) in a working paper stating that extensive and intensive training programs should tackle the barrier of inadequate technical and managerial skills for the implementation of energy efficiency improvements. According to Worrell et al. (2001), lack of information is likely to be even more of a barrier to EE measures in developing countries than in developed countries. First, the information infrastructure at the private and public levels tends to be less developed than in industrialized countries. For example, energy management systems or energy benchmarking are less pervasive in developing countries. Likewise, developing countries suffer from a limited public capacity for information dissemination, limited private technical capacity to access information (e.g. via internet), or lack of intermediary institutions providing information on the energy use of companies, processes, or technologies (e.g. via sector associations or company networks). Second, since acquiring and processing information depends on human capital infrastructure, companies in developing countries are less suited to effectively using existing information. Also, relevant information, for example, technology performance, may only be available in a foreign language (OECD, 2011). EE policies must address first the information barrier before additional policy interventions can be successful.

According to Worrell (2011), the policy implemented in the steel industry was first piloted by the China Sustainable Energy program in association with the Lawrence Berkeley National Laboratory. The agreement that focuses on industrial EE improvement will need to be negotiated. Typically, these agreements are contracts between industry and government involving a long-term commitment. Government participation in agreeing on such support will facilitate information dissemination, assessments, financial investments, and benchmarking, which are crucial in overcoming EE barriers.

According to OECD's 2015 report, the promotion of EnMS uptake includes a range of possible incentives due to the large variation in the market and regulatory forces in different countries. EnMS incentives can improve the participation of industries in implementing EE; however, it can also inadvertently lead to restricting private sector investment support. There are many possible hazards that can happen if incentives are unnecessarily used or are not well adapted to the real need of industry. The main barrier to EnMS uptake is a lack of knowledge rather than financial constraints.

A report by DOE (2015b) states that the lack of available knowledge within federal, state, and utility incentives for end-use EE measures can lead to missed opportunities. Another opportunity missed by consumers is due to the lack of sufficient in-house technical expertise and non-participation in demand response programs. Demand response is a program where consumers of electricity can take financial advantage by changing their electric usage from their normal consumption patterns in response to changes in the price of electricity over time, to provide incentive payments designed to induce lower electricity use at times of high wholesale market prices, or when system reliability is jeopardized. Lack of knowledge is a key barrier in adapting EE technologies leading to missed opportunities.

Technological obstacles in the Philippines are seen to be very low. There is sufficient technical expertise in the country, including professionals with previous experience of EE in large corporations, who would be able to adopt new and emerging technologies easily. There is insufficient knowledge across all sectors in the Philippines about the needs and benefits (e.g. financial savings) of implementing energy efficiency. This creates a lack of motivation for executives and chief executive officers (CEOs) to implement EE and to make an informed decision on EE investments. Knowledge and capacity regarding energy-efficient technologies in Vietnam are low, and there is also a scarcity of quality EE service providers in the country. In addition, Vietnam uses very low-grade coal (for example, anthracite) which prevents EE measures to be implemented on boilers and other combustion equipment (Copenhagen Centre on Energy Efficiency, 2015). There is a lack of targeted development programs for specific sectors and consumers, e.g. brick kilns, boilers, air conditioners, etc.

Implementation of Energy Efficiency Technologies. A report by IEA (2011) categorized a set of twenty-five policy recommendations to be provided for the seven priority areas. The priority areas are listed below:

- Cross-sectional
- Transport
- Buildings
- Industry
- Appliance and equipment
- Energy utilities
- Lighting

The twenty-five recommendations have received high-level political and stakeholder support and resulted in the increased implementation of EE. This section of the literature review focuses on only two of the above-mentioned priority areas on electrical consumers of commercial (buildings) and industrial sectors in how EE technologies can be implemented (IEA, 2011, p. 3).

Existing Implemented Models on Energy Efficiency Technologies. Economies industrialize the dependency on more sophisticated infrastructure, thus causing an increase in a technology system that makes energy more important. The world's continued industrialization and economic growth could be potentially threatening due to the number of energy-related problems and constraints (Fawkes et al., 2016). Most studies show that European Union countries are often responsible for the biggest part of energy consumption, especially targeting buildings which consume over 40% of total energy. It is found that an artificial intelligence model, such as neural networks and support vector machines where these technologies have a high potential for mapping environmental conditions in real situations, could be used to control energy consumption. A study was conducted to compare a neural network model that utilizes a statistical and analytical model with a model benefiting from the multi-objective algorithm. An experiment was conducted at the Solar Energy Research Centre (CIESOL in Spanish) located at the University of Almeria, Spain where neural network models are compared to a naïve autoregressive baseline which can predict electricity demand. Results of the experiment showed that the models obtained from the multi-objective genetic algorithm perform comparably to the model obtained through a statistical and analytical approach (Khosravani, Castilla, Ruano, & Ferreira, 2016). But the utilization of only 0.8% of data samples have lower statistical validity, and the above approach to energy consumption control will need further validation and testing before being accepted.

The United Nations Environment Programme (UNEP) (2009) reported on climate change and building and highlighted six major key messages. First, the building sector has the most potential of delivering significant and cost-effective GHG emission reductions. Second, countries that have not implemented EE technologies will not meet the targets on emission reduction. Third, proven policies, technologies, and knowledge are available to reduce GHG emissions. Fourth, the building industry is committed to taking action, and many countries are already playing leading roles in GHG emission reduction. Fifth, a key factor is that there are significant co-benefits, including employment, that will be created by policies encouraging EE and low GHG emission building activities. Finally, if the building sector fails in encouraging low carbon emission and EE in new building construction or in a retrofitted building, it will lock countries into a disadvantage of poor performance buildings for decades. Among the six key messages mentioned above, the four that have been prioritized are first being the national emission targets in reducing GHG emission in the building sector. Second, in order to prove that the GHG reduction goals are supported, buildings that have implemented a reduction in GHG emission have to be certified by Nationally Appropriate Mitigating Action (NAMA). Third, under the Kyoto Protocol (UNFCCC, 1998), the Clean Development Mechanism (CDM) (UNFCCC, 2006) was developed where the process can be implemented in supporting EE program for buildings. Finally, GHG emissions data can be developed in maintaining baselines for consistent approach on reporting and monitoring on performance (Eang, 2015). The issue here is the proving of GHG emission reduction, which is difficult to establish as energy usage is very dynamic, utilizes various fuel sources, and the CDM process to support EE programs has since faded. The aspiration of UNDP will not be met unless there is a measurement and validation protocol that is accepted internationally to measure GHG reduction and an alternative mechanism to replace CDM. Under the United Nations Framework for Conventions on Climate Change (UNFCCC) found in the Kyoto Protocol, the first commitment period by participating nations was from 2008 to 2012. Under the Doha Amendment, the second commitment period is from 1st January 2012 to 31st December 2020 (UNFCCC, 2012).

Thollander, Sa, Paramonova, & Cagno (2015) investigated energy efficiency opportunities and energy management practices in the foundry industry. Foundry is an energy-intensive industry, and energy accounting is necessary to determine where and how energy is being consumed and the efficiency of the energy management system. It was found that foundry industries consume more energy than any type of industry. An energy accounting method should define areas of higher energy use, wasted energy, and energy savings potential. To determine energy consumption patterns, an energy audit is the main part of the process. Energy consumption patterns can help in understanding the way energy is being utilized and help to control energy cost by identifying energy wastages and areas of possible energy savings. Energy management is very important as it deals with adjusting and optimizing energy demands by utilizing EE technologies and procedures (Thollander, et al., 2015). The study by Arasu and Jeffrey (2009) provides a case study of the energy required to produce one ton of liquid metal in four foundries. This provides an idea of the current energy consumption of the foundries that can be compared to standard norms and can be used to implement deviation control methods. There have been only a few studies carried out to establish the industry's baselines in energy consumption. There should be more studies established which will cover all sectors of energy consumers-based climate conditions, which will provide a realistic baseline to establish EE measures.

According to the 2017 European Commission report, it is important to provide an effective management process for policymakers since they control industrial EE policy. Full knowledge and guidance on understanding energy management can pave the way for effective EE implementation. Policymakers should also make a resolution about any industry efficiency policy and check if the policies are well proven and tested. The documents on the implementation of the EE process must be written in a manner where non-technical readers can easily understand. All key factors on energy management should be communicated to all departments in an organization, and energy management procedures must be followed with discipline and made as a standard for all types of industry. This process was actually implemented in 1973 and 1979 during the 1970s oil crisis. The implementation of EE targeted the building industry without affecting its performance as energy managers found the significance of energy prices over these periods. Efficient management in implementing EE processes in an organization can achieve full control of energy usage. Management has to resolve high energy cost through the implementation of cost-effective energy technologies and to overcome technical issues that arise from these implementations.

There are recommended features for an efficient energy management system in this study. First, the energy management procedural document should be consistent and written for easy understanding. Second, the energy management plan or process should be distributed across the entire firm and not just to the technical and engineering departments. Apart from this, management should provide support and make resources accessible to the entire organization. When these are properly scheduled and configured, it can create a situation where an assumption can be proven to be real. Always consider the pertinent performance measurements and feedback from the employees as it can help with EE performance. Using a unique life-cycle approach can help in making decisions if a new plant or equipment is required. Organizations should carry out continual improvement on their EE programs where their objectives are attainable and improvements are long term with the performance measured periodical (European Commission, 2017). The main barrier here will be the cost of implementing and maintaining an energy management system. Large revenue earning organizations will be able to afford the cost of implementing and managing energy management systems. What, though, can be done about smaller organizations that are unable to afford the implementation of an energy management system, let alone manage it? There must be some sort of financial scheme available to overcome the above-mentioned barrier for smaller organizations.

The Use of Energy Control Systems. According to Capgemini (2011), the strengthening of existing control systems and utilities can help in the reduction of energy consumption by operating closer to the level of energy that the organization has targeted. Small investments are necessary to identify the new efficient operating parameter; to implement a control mechanism; and to repair, reinstate, or to replace the existing system. There are many types of systems that fall into the category of the control system. Some are based on a primary requirement and are of simple design. There are fields of possible implementation. An example is to control the overuse of energy for an air-conditioning system which measures a pertinent operating parameter to control or balance it. Synchronizing the energy demand to the supply and carrying out recommended predictive maintenance for system or equipment will save energy. Continuous information on water consumption, with the correct control strategy, can assist in reducing excess water flow. The quality of a product is dependent on its efficiency and life span, and if a product's life cycle is known, then planned replacement of the product could be carried out before it fails. Using speed drives and control loop tuning should also help in monitoring the performance of each key plant processors. If there has to be the highest form of maturity change, then the basic design of the model's processes has to be changed to improve the design concept in order to reduce energy supply. According to the 2017 European Commission report, energy maturity matrix can give the biggest energy savings since it provides a high-level assessment of strengths and weaknesses across six areas of energy management. These areas include energy policy, organizing, training, performance measurement, communication, and investment. Examples of an organization that implements energy maturity matrix are power and heat plants. The process lines should correctly synchronize with process equipment where the efficient operations of cooling or district heating can optimize energy consumption. However, to implement the energy control system and energy maturity matrix, it is necessary to expect more financing and risk when compared to their other EE projects.

From the report provided by IEA (2016a), the implementation of EE technologies can be done in two ways. First, the specialist of an organization must be aware and must have knowledge on how the system can be monitored and be able to identify what can be done to apply EE in a specific area. Second, the energy consuming equipment installed and operated within the organization must have the potential for energy reduction or energy efficiency improvement, and there should be available EE systems or technologies to support the reduction or improvement respectively. However, these are not easy tasks to be carried out as the right trained experienced technical personnel are required to identify and analyze energy savings potentials. These types of technical personnel are scarce and are highly sought after, so an organization should find an alternative to what has been recommended above, such as seeking EE experts outside the organization.

Standardizing Energy Management System. ISO 50001 is a voluntary international standard for energy management system (EnMS) and was developed by the International Organization for Standardization (ISO, 2011) in collaboration with the United Nation Industrial Development Organization (UNIDO) as a response to climate change and the need to have a standard global energy management system. The standard was developed to equip organizations with the requirement of an EnMS that derives its standard from national and regional energy management standards, specifications, and regulations. The key focus of EnMS is that it involves all levels and functions of the company and requires the participation of top management. This ensures the continuous motivation of employees, a pertinent element in the effective functioning of the EnMS framework. The ISO 50001 can be incorporated into the existing management structure of an organization as an organization's policy or part of its strategic objectives. The implementation of ISO 50001 is an effective means of overcoming the prevalent informational, institutional, and behavioral barriers to EE (OECD, 2015).

Implementation of an energy management standard within an organization requires changes in existing institutional practices toward energy, a process that could benefit from technical assistance from experts outside the organization. An organization's staff, familiar with management systems like quality, safety, and environment, understands the dynamics of establishing a management system and its successful integration into the organization's corporate culture. These experts, however, typically have little or no expertise in EE. In contrast, industrial EE experts are highly specialized in EE but are trained and oriented toward the identification and execution of energy efficiency projects without a management system context. The appropriate application of energy management standards requires significant training and skill. There is a need to build not only an internal capacity within the organizations seeking to apply the standard but also an external capacity from knowledgeable experts to help establish an effective implementation structure (Huang, 2011). The suite of skills required to provide the technical assistance needed for energy management is unique since it combines both management systems and EE.

Whether an energy management system implemented in an organization is customized or is of ISO 50001 standards, the basic EnMS process is based on the "Plan-Do-Check-Act", continual improvement framework illustrated in Figure 7 and described below:

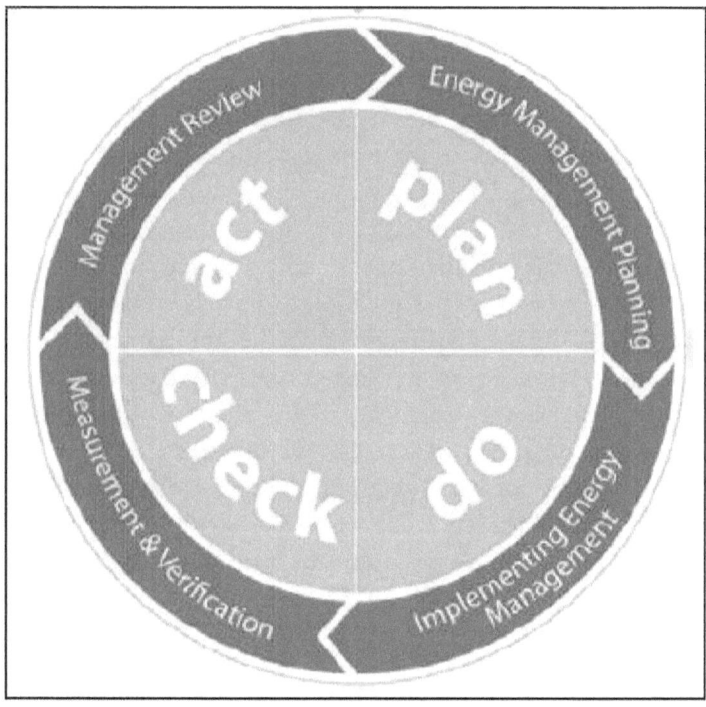

Figure 7 EnMS process. Adapted from "Energy Efficiency in the Steel Sector: Why It Works Well, But Not Always", by OECD (2015).

In the Act, the plan consists of conducting an energy review, establishing baselines, and utilizing these baselines to benchmark against similar sites, set objectives, and target to develop resources with the organization's energy policy. "Do" is to continuously evaluate the action plans formulated based on the targets created during the Plan phase. The Check phase is to continuously monitor and measure processes and to review the level of target achievement and the effectiveness of the EnMS against the objective of the energy derived in the Plan phase. The next is the Act phase where the results are published, and positive results will be rewarded. New objectives are derived by repeating the "Plan-Do-Check-Act" cycle on a periodic basis, resulting in continually improved energy management performance (OECD, 2015). As mentioned previously, the cost to implement and manage an EnMS system is very costly and not affordable by all.

According to McKane, Scheihing, and William (2007), EE in industrial sectors is usually achieved by changing the operation of an industrial site utilizing Best Available Technologies (BATs). The first step in identifying and prioritizing the full range of EE opportunities is by implementing an energy management system in advance of other major investment decisions related to physical capital, while stand-alone BAT investments are still important in achieving the full potential of EE. It has been shown that experienced industrial companies can save up to 10–30% of their annual energy consumption and a similar reduction in their operating costs through better energy management involving operational changes alone. Many of these energy-saving opportunities can also be achieved by applying proven best practices, with short paybacks of one to two years or, in some cases, within months. The overall monitoring of system efficiencies, optimization of industrial systems, and improvement in energy performance in a company are achieved through the capabilities of the EnMS (OECD, 2015). The other benefits in terms of productivity for companies using an EnMS can also include enhanced production and capacity utilization, reduced resource use and pollution, and lower operation and maintenance (O&M) costs. All of these results increase the value and competitiveness for the company, but the downside is the costliness of implementing and managing it.

According to a technical report by Tanaka, Watanabe, and Endou (2010), there are various types of energy-related information in a plant acquired by an online execution system that are presented to the end-user through user interfaces. Parties concerned with energy are expanding to every hierarchical level within an organization due to growing concerns about high energy consumption, depletion of natural resources, global warming, etc. The users of conventional energy management systems are supposed to be only managers of energy supplying utility equipment. However, the Enerize E3, a factory energy management system developed by Yokogawa, Japan, is the first system in the industry to standardize energy management, encouraging all members in a factory to participate in energy-saving activities. It assumes a wider variety of people as its users, and hence, such a model system might act as a basis for implementing energy management systems in the Klang Valley region. They include personnel concerned with production control or production engineering, personnel belonging to the general affairs department concerned with plant layouts, and personnel belonging to corporate social responsibility (CSR) departments. The issue with these kinds of systems is that they are customized systems that can only be applied to a unique plant and cannot be adopted as a general solution across the industries as each industry's processes differ. One way of overcoming this barrier, even though a complete EE solution from a particular industry cannot be utilized directly by another industry, is to identify sub-systems of the complete EE solution that can be applicable to the other industry, which might provide a proportion of the energy reduction (Chai & Yeo, 2012).

Reports of Energy Efficiency Technologies. A study by Li et al. (2011) reported that at the AT&T Labs-Research in Red Bank, New Jersey, USA, the basic concept of energy-efficient communications are first introduced, and then existing fundamental works and advanced techniques for energy efficiency are summarized. From these basics, advanced techniques related to EE pave the way for a sustainable future. These can include Orthogonal Frequency Division Multiple Access (OFDMA) networks, information-theoretic analysis, resources allocation, and relay transmission. Some important benchmarks on energy-efficient design related to future inventions are briefly discussed in this section.

There is a significant amount of scope when it comes to the implementation and design with energy efficiency technologies and large numbers of effective outcomes when it comes to the significance of EE policies. The best practices that are related to it will be basically developed by policies to explore, motivate, and share its experiences (United Nations Economic Commission for Europe, 2017). With the ample amount of information that is available on energy efficiency, it is mandatory that sales within energy efficiency systems have an expressible baseline (EC-Europa, 2017), and savings from EE implementation should be used to achieve that (World Energy Council, 2013).

Some EE technical solutions are already available for implementation and some of them are already in use. Solutions to technical barriers can indirectly overcome the barriers associated with cost-effectiveness, acceptance, environmental impact, innovation, and financing in the implementation of EE technologies. Governmental support can also promote its contributions by investing in EE technologies (EPA, 2017). If an organization or government intends to build on EE, a long-term commitment is required, and the implementation process could take a long time before it shows the effectiveness of the EE technologies. Sometimes the rebound effect might only appear toward the latter stage of an EE project. Willingness to accept and utilize new technologies is essential in spite of successful implementation. Environmental assessment is normally based on achievable targets, which can be realistic and achievable. Decision makers will also need to learn about useful tools available for further innovations and future development in energy efficiency technology. Understanding these, and the barriers associated with it, will help in the process of energy efficiency technologies implementation (Cagno, Worrell, Trianni, & Pugliese, 2013).

Energy production gains are dependent on the level of technologies and the spread of efficient processes. There is still significant potential in implementing EE economically since the usage of EE does exist. Each year the total Gross Domestic Product (GDP) amount has decreased by 1.4 percent since 1990 (Crafts, 2004). The group of 20 nations (G20) that represent 85% of the global economies holds its annual summit meetings to discuss and address global crises such as GHG and CO_2 emissions. A practical Energy Efficiency Action Plan (EEAP) to strengthen voluntary EE collaboration in a flexible way was adopted by the G20 members in 2014 based on the activities that best reflect their domestic priorities and interests. It allowed countries, through an opt-in basis, to share their knowledge, experiences, and resources. The work conducted by participating members under the 2014 EEAP constitutes the foundation for G20 collaborative action on EE. The Energy Efficiency Leading Programme (EELP), apart from covering the existing activities under the EEAP on vehicles, networked devices, finance, buildings, industrial processes, and electricity generation, expands its work areas to include five new key areas of collaboration: Super-efficient Equipment and Appliances Deployment (SEAD) initiative, Best Available Technologies and Practices (TOP TENs), District Energy Systems (DES), Energy Efficiency Knowledge Sharing Framework, and Energy End-Use-Data and Energy Efficiency Metrics (IPEEC, 2016).

Applied EE Metric. In 2009, the report of calculating energy usages was measured at parity with primary energy density. Primary energy intensity is considered a total energy equivalent to one unit of GDP (EEA, 2017). Power parity of energy usage is found more in the Commonwealth of Independent States (CIS) compared to European countries. It has been recorded that energy usage in the Organization for Economic Co-operation and Development (OECD), Asia, and Latin America goes beyond 15 percent, while North America is the highest standing country for energy utilization. India and other countries in Asia give a parallel record with the world's average, reaching above 60 percent when compared to European countries. The countries utilizing higher EE are CIS, the Middle East, and China (ABB, 2017). The high utilization is due to various factors that highlight the development of energy-intensive industries and the increase of energy-related prices. The average value of EE in surrounding Asian countries is found to exceed more than the average value of total average energy consumption. After such records or discussions, it is found that only Latin America and Middle East countries have positively improved in their EE system. Other countries that have targeted to reduce energy consumption show negative values. These values were calculated in 2009. Except for the Middle East, the intensity of global energy had decreased by 1.4 percent between 1990 and 2009. The trend toward higher energy prices had a combination effect on EE programs such as CO_2 abatement policies and economic service activities in OECD countries. In 1990, the regions or countries that recorded the highest primary intensity seemed to have reduced. India, CIS, and China recorded the highest drop, and China seems to show the largest drop among all these countries. India and CIS countries' reduction rates are slower than that of China's since its reduction is twice that of India and CIS. The rapid growth of machines and transport sectors were found to be the major factor for the reduction in energy growth. The attributes found to be more effective were the efficient use of coal and the management of higher energy prices. The above average group was found in North America and the European Union (EU). In addition, it was found that there has been a significant change in the percentages of energy usage in these areas. The implementation of EE programs is found to be uncertain as the continual use of EE technology does not actually give the same results as before.

There are many strategies that can be mentioned with respect to action plans in energy efficiency implementation that require clear objectives, timelines, and established methods. A regularly updated strategy and action plan for improving energy efficiency throughout a country's domestic market should be formulated based on an analysis of energy use, markets, technologies, and efficiency opportunities. An EE project that is found cost-effective must have actions and strategies to minimize, remove, and overcome known barriers before implementation. Providing assessment opportunities for energy efficiency improvement can be done by prioritizing areas with government policies and guidelines that are likely to yield the most cost-effective and largest improvement (IEA, 2011). In developing energy efficiency strategies and plans, there should be coherence between energy, environment, climate, and the economy. The experiences and analysis carried out by other countries must be taken into account while formulating strategies and action plans when implementing energy efficiency measures. Strategies and action plans for new and emerging technologies should have continuous integration and coordination. Periodically, there should be a review of regulation and subsidies to ensure that the retail energy prices are reflective of the full cost of energy supply, delivery, and inclusive of environmental cost. The implementation of energy efficiency, however, can be started from private investments targeting government facilities. From this energy efficiency implementation, verification protocols, standardization, energy efficiency capacity building, and other protocols could be reached through demonstration and the publication of the energy efficiency measures outcome. The dissemination and generation of knowledge with reliable technical assistance will be viewed with the opportunities needed on the basis of energy advisory services. Training programs and education will ensure that all of the sectors are aware of energy efficiency technologies (IEA, 2011). The verification and measurement protocols that are initiated to quantify the benefits of EE technologies implementation could encourage others to follow.

There are many uncertainties pertaining to EE verification protocols and measurement tools, and the insurance of overcoming this is in the investment in EE technologies. Collaboration is needed with private financial institutions to develop a public-private partnership and the other frameworks that facilitate EE implementation by means of EE financing. EE policies should be periodically monitored, enforced, evaluated, and updated where policy and program effectiveness should be evaluated during and after implementation, and non-compliance should be identified, reported, and made public. The process of identifying non-compliance should be fair and transparent, and the associated penalties should be clear and serve as a constructive detriment to non-compliances (IEA, 2011).

In order to scale-up investment in EE, there needs to be a policy-driven framework which is primarily initiated by the local government and demand-driven model usually established by the Energy Services Companies (ESCOs). A policy-driven framework to support the scale-up of EE investments that the local government can utilize are market-oriented incentive schemes, implementing standard and instrument for EE, and partnership arrangement with the private sectors to increase market certainty. The demand-driven model can be created by ESCOs that can provide the full range of services, including design, implementation, and financing for an energy efficiency project. Large upfront cost (mainly for retrofits) still needs to be integrated into a business model when demand for EE increases. These costs could be covered through energy savings by innovative financing in a form of performance contracting. This method offsets the EE investment cost against energy savings across the financing term, providing a zero-net-cost investment technique. ESCOs could also leverage where development banks establish special purpose vehicles for buying matured loan from them (Bache, 2014).

Every large-scale industry has to support and invest in research and development (R&D) in order to increase the implementation of EE through new technology innovations that have been demonstrated and proven. Incentives for utilizing EE technologies and imposing penalties for not using EE technologies can promote the implementation of EE. Promoting awareness with local and international organizations on further corporations and collaboration can be a positive step in the encouragement of EE implementation. However, the rate of implementation of EE is dependent on whether the total cost of the EE project, including capital transaction, administration, operations, and maintenance cost, is recovered from the savings in an acceptable payback period of the investment. In addition, it needs to be continuous to provide savings after this payback period. If the cost of EE technologies is high, the longer the payback period will be for the investment and the need to import more fossil, which will soon have a drastic effect on its pricing. A proven and sustainable EE technology could reduce the consumption of fossil fuel, resulting in benefiting a country's trade balance as it reduces the import of fossil fuel (OECD, 2017). The actual target of these practices is to increase the energy efficiency of energy consuming systems where the input to an energy consuming system will be reduced to produce the same output, and if these practices are not carried out, it can negatively impact the environment. The term EE varies from each person and the context of actually applying it can be used in different ways. The general definition of energy efficiency refers to the amount of input energy compared to the output equivalent energy for engineers, equipment, and machines (Noka Group, 2017). For example, the EE in an electric motor is defined by the ratio of electrical output and input values.

CRITIQUE OF PREVIOUS RESEARCH

The following is a summary of previous research supporting this study's research questions. The previous studies were all well-written critiques that employed a qualitative approach, had excellent presentation of data, provided clear themes, and described how EE had been successfully implemented by overcoming barriers. The previous studies conducted that were excluded were due to the following weaknesses: lack of peer debriefing or analysis corroboration, insufficient information to determine transferability, an apparent role conflict by the researcher, and insufficient descriptions of EE barriers. Critique was also carried out on general adaptability related models where the existing theoretical orientation needed improvement.

Usefulness of Implementing Energy-Efficient Technologies for the Consumers. The implementation of EE measures at all stages of the supply and demand chain could significantly reduce the negative impact of energy use on the environment and human well-being and increase the availability of primary energy reserves while achieving maximum benefits in terms of outputs from the available energy. The cost to both suppliers and consumers can be reduced while maintaining the same level of energy-dependent activities. We have seen how the overall efficiency of an energy-dependent activity, from the primary energy resource to the final output, could be improved. It is a figure that represents the cumulative effect of all the inefficiencies along the supply-demand chain. Creating and implementing measures to improve EE at each step will contribute to increasing the final figure, so even small improvements can have a significant impact. To improve demand, EE will clearly reduce energy costs directly to the energy end-users, thus improve generating, transmission, and distribution efficiencies. Supply-side actions carried out by utility companies can also bring cost benefits to the end-users in the typical regulatory environment by ensuring energy prices are well controlled. Indeed, the combined effect of supply and demand-side EE improvement means that the load on generating facilities is lowered, and this can help keep older systems and equipment in good condition. This is because lower overall loads often allow the equipment to run below maximum capacity or be shut down more frequently (or for longer periods) for preventive maintenance. Older equipment will usually need more maintenance. Depending on system characteristics, forced shutdowns for repair can be reduced and system operating efficiency can be raised; overall system reliability can be improved as a result. Improving energy management is almost always a low-cost action that achieves valuable benefits in the short term. Maintaining good management ensures these benefits are continually contributing to enterprise profits (and the national economy) in the long term (UNIDO, 2017).

Economic gains related to cost reductions resulting from lowered energy use and threats of rising energy prices are the most important drivers for implementing EE measures or technologies. In addition, government efficiency requirements are another important promoting factor for EE implementation. An essential extension of this research work will be to incorporate the views of external stakeholders like researchers, equipment dealers, financial organizations, local government, trade associations/unions, and many more who play a role in eradicating barriers and are drivers for improving industrial EE. By so doing, claims made by respondents could be supported or refuted, and an additional broad base knowledge suitable for policy implementation will be developed. Lastly, studies could be narrowed down to high energy intensive and low energy intensive firms (Apeaning, 2012).

EE offers many cost-effective opportunities to achieve energy security to improve business productivity and to mitigate GHG. Industries' energy conservation measures make business sense since in almost all cases the payback period is less than five years. Also, needing to be factored into the equation are the rising fuel prices which augment savings in the long run. Basically, some EE measures such as replacement of single glazing with double glazing, provision of insulation in roof or utilization of high reflective paints on roof exposed surface, replacement of existing lighting system with Light Emitting Diode (LED) lighting, and replacement of Constant Speed Drive (CSD) chillers with Variable Speed Drive (VSD) chillers with the installation of plant optimizer are recommended (TERI, 2013).

Increased Gain on Markets. Globally, EE has the capability to reduce GHG reduction by approximately 40 percent at a cost of less than €60 per Metric Ton of carbon dioxide equivalent (tCO_2e). It has an upfront investment with an attractive payback period that provides added benefits by reducing the cost of energy and increasing the energy productivity of the economy. Many governments have emphasized EE opportunities during the economic downturns as a way to stimulate economies, create jobs, reduce domestic dependence on foreign energy supplies, and most importantly, to reduce carbon emission related to energy usage (McKinsey on Society, 2010).

There are drastic changes to be made if every facet is to be used in reducing the consumption of energy. The demand on producing more power is an inevitable one and the power that is needed for satisfying and balancing energy consumption must be ensured throughout the supply chain. If more energy needs to be produced, then accelerating the power together with implementing EE technologies can be a smarter choice since the cost and the total energy consumed can be decreased (Schock & Sims, 2012).

Investment in EE will follow suit with the expansion of policies. The global investment in EE was USD221 billion in 2015; this is an increase of six percent compared to 2014 as estimated by the International Energy Agency (IEA, 2016a). In 2015, the investment in EE was two-thirds greater than the investment in conventional power generation. The strongest investment in EE was from the building sector, contributing nine percent, with the United States investing close to a quarter of all efficiency investment in this sector. The largest investor in EE vehicle market with 41% of efficient vehicle investment worldwide was China. The EE services market is a sizable and distinctive market with a turnover of USD24 billion made by ESCO in delivering EE solutions. In 2015, China's ESCOs employed over 600,000 people with a revenue growth of seven percent, making it the largest market for ESCOs. In the United States, the ESCOs revenues in 2015 was USD 6.4 billion, which is more than double compared to the last ten years (IEA, 2016b).

The evidence indicates that the EE market will grow in the coming years. There is a new market entrance due to the increase in utilities, technology providers, and energy equipment manufacturers merging and acquiring ESCOs. Traditional energy utilities are providing energy services as a way to expand their revenue in the light of the low energy demand outlook in the IEA member countries. Development of new business models and service solutions are possible due to the growth of remote monitoring, control, and data analytics. EE companies are intending to gain a competitive advantage by improving the utilization of products based on prices. Many utility regulations are based on factors relating to higher operating efficiency within energy producers. Many reduced environmental impacts will be viewed if it serves as a major tool for efficiency implications on purchasing decisions. Companies that are intending to go green factor in environmental benefits like reduced local pollution. Every company's output will be increased based on the significance of improvements that are resolved with reduced pollution emissions (UNIDO, 2017).

The major advantage of applying advanced energy-efficient components in building construction has mostly to do with minimizing thermal bridges that reduce heat loss or gain and reduce risks of drafts and consequently deliver better indoor comfort for occupants, apart from saving energy.

Increase in Energy Efficiency Technology Usage Leading to Economic Development. Energy policies that are applied in the industry are promoting to provide low carbon and affordable as well as secure services. Fundamental assessment on the role of availability on efficient conversion is depended on the energy sources that value useful work on economic development. Increasing availability and efficient conversion decrease the input of energy that plays a key role in building economic growth (Ayres & Warr, 2010). These perspectives have some major effects on economic growth, which may limit the projections based on the turning point of easily affordable prices with respect to the imposition of higher carbon prices.

This perspective has serious implications for economic growth projections, since we may be at a turning point in affordable and easily available energy. Fossil fuel sources are becoming increasingly expensive due to the depletion of easily accessible reserves, the imposition of higher carbon prices, and the generation from low carbon energy sources is still relatively expensive. Thus, the continuation of past trends in economic growth rates cannot be assumed. This paper applies a co-evolutionary framework (Foxon, 2011) to raise the issue of the contribution of the efficient conversion of energy sources to long-term economic growth, to examine positive feedbacks or virtuous cycles between changes in energy input, and the conversion costs and changes in economic activity. This analysis highlights the importance of both EE improvements and changes in end-use consumption patterns to ensuring continuing economic prosperity as countries undergo a transition to low carbon energy systems (Foxon & Steinberger, 2013).

The implications beyond the employment are beyond the scope of this project. In the past, the reduction in employment caused by substituting energy for labor has been offset by new jobs created through the growth of the economy as a whole. For ecological economists arguing that increasing GDP growth should no longer be the main aim of national economic policy, this implies that, as well as higher investment in low carbon and EE innovation is required, investment needs to be reoriented toward economy sectors that are more labor intensive, such as caring for children and older people. These sectors are less productive in conventional economic terms (Foxon & Steinberger, 2013).

According to Carley and Lawrence (2014), Energy-Based Economic Development (EBED) offers researchers and practitioners an integrated framework to align with what are often considered disparate goals into a unified approach. Finally, EBED recognizes the role of governance, leadership, and stakeholder models in shaping its success. EBED is a direct extension of the energy planning and economic development disciplines. As both founding disciplines have evolved through the years, a number of synergies in practice and objectives have emerged marking the nexus of the EBED domain.

High-technology industries provide opportunities for economic growth, but this also raises concerns because of their energy demanding nature. A nation-wide industrial input-output analysis has been conducted to demonstrate the positive effects of science parks on national economic developments and industrial upgrades. The concept of energy intensity and an energy-efficient economy index are applied to an integrated assessment of the relationship between economic growth and energy consumption. The proposed case study suggests that economic and EE objectives can be simultaneously achieved by the development of high-technology industries, while three energy policy implications are considered. First, a macro viewpoint for both national and international scenario is needed, and high-technology industries should be considered as parts of the national and regional economies by governmental agencies. Second, a proper industrial clustering mechanism and a shared environmental facility supported by the government, such as planned land and road usage, electricity and water supply, telecommunications system, sewerage system, and wastewater treatments, can improve EE for high-technology industries. Third, governmental policies on the taxing and management system in science parks would also direct the energy-efficient economy of high-technology industries (Yan & Chien, 2013).

Science and technology policies and high-technology industrial developments can bring about significant national economic development. At the same time, the energy demanding nature of high-technology industries is one of the major concerns of public policymakers and relevant stakeholders. Hence, conducting an integrated evaluation of energy use and economic development to accept the philosophy of a low carbon economy rather than impede the industrial developments has become one of the most pragmatic low carbon energy-saving implementation approaches. For the governmentally supported science parks, the proposed integrated analysis and performance evaluation provide a more comprehensive perspective in supporting the performance evaluation of industrial incubation policies. From a more macro standpoint, the integrated economic/energy perspective should be included in the list of those factors used for the construction of nation-wide policies on industry choices, segmentations, and structural developments. The results of the present study suggest that the national science park's policy can facilitate the successful development of high-technology industries, which bring about both significant economic benefits and EE performance. Thus, high-technology industries need not necessarily be considered as high energy consuming based simply on absolute environmental standards as the sole perspective. An integrated evaluation of the impacts of high-technology industries should be conducted to obtain more comprehensive policy directions. While the operations of the science parks in Taiwan have long enjoyed governmental support, the proposed study provides several energy policy implications (Yan & Chien, 2013).

General Adaptability Related Model for the Adoption of Energy-Efficient Technologies. The general models related to the adaptation of applied EE technologies are discussed below. The main barriers considered in this research are categorized as follows: Group 1 policies and regulatory; Group 2 economic, finance, and market; and Group 3 behavior, informational, and technical. The models that can be applied for this research are shown below:

Core Competencies Framework

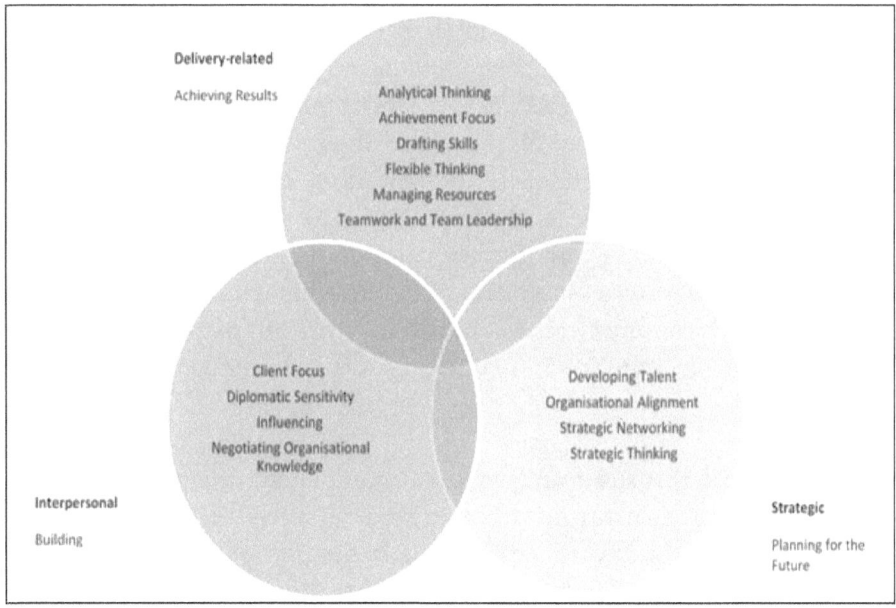

Figure 8 Core competencies framework. Adapted from "Competency Framework" (2014).

The core competency framework developed and recommend by OECD (see Figure 8) could be utilized as the framework for this research as shown in the blue cluster, e.g. interpersonal relationships competency components are required in continuously analyzing and adapting changes and requirements to the organization's advantage, which could be drivers to overcome Group 1 barriers identified in this research. The green cluster is the competency components required for the strategic planning for the future and are the core competencies required as drivers to overcome Group 2 barriers in this research in order to develop internal organization's policies such as EE policies, budget, and EE management system. The Group 3 barriers in this research could be overcome with the core competency components in the purple cluster as the driver for the smooth implementation of EE technologies (OECD, 2014).

In order to fill the research gaps, the current OECD core competency framework will require the following improvements.

Interpersonal. Under this sub-framework, the component of communication requires to be added in order to have an effective EE program for an organization. The element of communication will help strengthen the relationship between the internally and externally parties of an organization involved in EE. The internal client should be the persons responsible for EE of an organization, and the external client should be the regulatory authorities and any other related external organizations overseeing the implementation of the EE. In order to have a continuous and successful EE implementation, there should a coordinated communication between the internal and external clients so that the minimum mandatory EE requirements are fulfilled (Armel, 2014).

Strategic. Under this sub-framework, the component of evaluation has to be incorporated for an organization to have an effective EE program. In order to plan for the future, continuous evaluation of the organization's energy consumption trends and the market prevailing conditions on energy should be studied to ascertain the needs of EE implementation. A continuous feedback mechanism should be developed to ensure the energy security for an organization is continuously monitored (OECD, 2017).

Delivery Related. Under this sub-framework, the component of project management and EnMS needs to be incorporated to ensure effective delivery of EE measures of an organization. Proper project management will ensure EE projects meet the planned implementation time and budget (IEA, 2013).

Camden's Behaviors Framework. This framework consists of ten types of core behavior values (see Figure 9) where some will act as the main driver to overcome EE barriers, whereas the other core behavior values will act as a subset of the main drivers to overcome EE barriers.

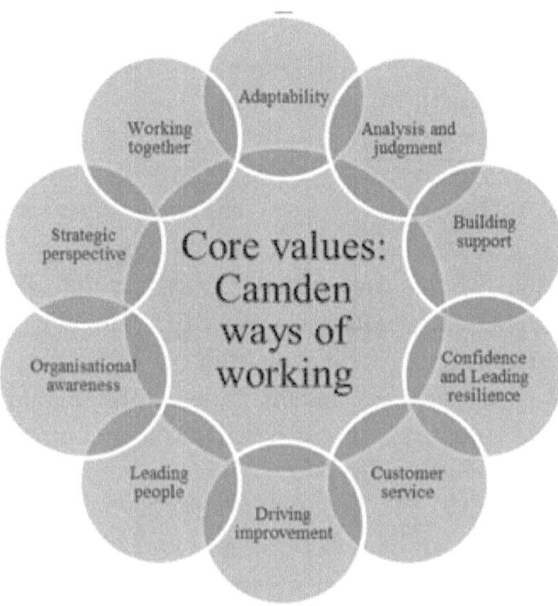

Figure 9 Camden's behaviors framework. Adapted from "Camden's behaviors framework Staff guide", by Camden (2012).

The Group 1 barriers identified in this research could be overcome by a combination of the above core behavioral values consisting of adaptability, analysis and judgment, building support, confidence and leading resilience, strategic perspective, and working together. The Group 2 barrier identified in this research could be overcome by utilizing core behaviors of this framework as drivers which consist of adaptability, analysis and judgment, building support, confidence and leading resilience, driving improvement and working together. The core behavior comprising of adaptability, customer service, driving improvement, leading people, organization awareness, and working together will be drivers to overcome Group 3 barriers of this research (Camden, 2012).

The additional behavioral values required to improve the current Camden's behaviors framework in order to meet the research gaps of this study are as follows:

Analysis and Judgment. When an EE barrier is analyzed and a judgment is taken, there needs to be another core value. The other core value is solution generation which by providing primary solutions together with alternative solutions will tremendously help in offering a roadmap toward the implementation of EE measures (Kadam, 2014).

Customer Service. For an organization to have an effective EE program, customer services should have two components, internal and external customer service. The internal customer should be a person or a department who is responsible for an organization's EE activities, and the external customer should be the regulatory authorities and any related external organizations overseeing the implementation of the EE. In order to have a continuous and successful EE implementation, there should a coordinated liaison between the internal and external clients to meet the minimum mandatory EE requirement (UNIDO, 2017).

Driving Improvement. For an organization to have an effective EE program, driving improvement alone will not necessarily meet the goals and objectives of an EE project. There needs to be another core value in terms of planning. Any improvement needs to be planned before being driven in order for an organization to realize the tangible and intangible benefits of an EE project (DOE, 2015b).

Leading People. In order for an organization to have an effective EE program, it needs to have the element of training. Before organization personnel can be led, e.g. to implement EE measures or project, adequate training has to be provided in order for the personnel to have an understanding of what is required of them to deliver before being led or being assigned the task (Kadam, 2014).

Competency Framework

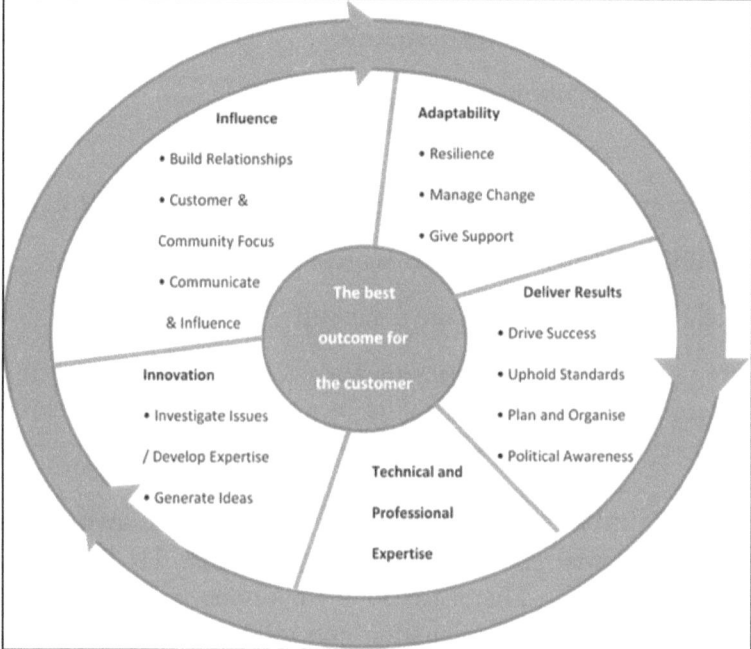

Figure 10 Leadership competencies framework. Adapted from "Leadership Competencies Framework", by Enfield Council (2017).

The competency values (see Figure 10) from this framework, consisting of adaptability, delivering results, and influence, could be drivers to overcome the barrier identified in Group 1 of this research. Group 2 barriers identified in this research could be overcome with competency characteristics consisting of adaptability, deliver results, innovation, and technical and professional expertise from this framework as drivers. The Group 3 barrier identified in this research could be overcome with competency drivers consisting of adaptability, delivering results, influence, and technical and professional expertise (New Enfield, 2017).

The additional competency values required to improve the current competency framework by New Enfield to meet the research gaps of this study are to clearly define the technical and professional expertise requirements. To have an effective EE program in an organization, a leader should have the ability to identify among the current organization's in-house technical and professional personnel if they have the following capabilities:

- Ability to identify EE opportunities
- Identify EE solutions
- Analyze financial feasibility of EE solutions
- Ability to plan, deliver, install, test, and commission EE solution
- Ability to set up an EnMS to manage EE solutions
- Ability to train others in the organization in EE and EnMS

If the above-mentioned capabilities are not available within the organization, then the leader should have the ability to source for external technical and professional expertise (UNIDO, 2017).

Motivation-Capability-Implementation-Results Framework (MCIR). MCIR is a conceptual generic framework (see Figure 11) based on stage-wise process feedback. There are four important stages comprising of motivation, capability, implementation and results for the adaptation, and implementation of EE practices as a process with a feedback effect. Factors affecting EE adoption and the interest and objective of stakeholders are reflected at each stage. The barriers identified in Group 1 and 2 of this research will have a direct bearing in the motivation stage. At the capability stage, barriers identified in Group 2 and 3 will have an influence. In the implementation stage, the barriers primarily identified in Group 3 will have an effect. Finally, at the results stage, the barriers identified will have a bearing on the outcome of the EE practices' success.

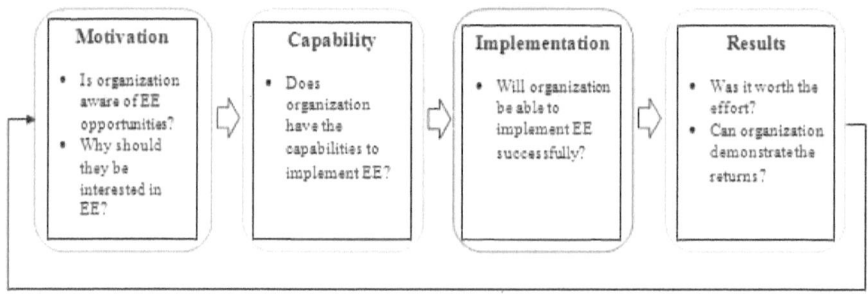

Figure 11 Motivation-Capability-Implementation-Results (MCIR) framework. Adapted from "Overcoming energy efficiency barriers through systems approach—A conceptual framework", by Chai & Yeo (2012).

The key component missing from the current MCIR framework to meet the research gaps of this study is an opportunity. Without having EE opportunities in an organization, the application of the MCIR framework will be totally ineffective. Therefore, identifying EE opportunities requires expert skill and experience and so an organization should have the required personnel who has the following capabilities:

- Identify EE opportunities
- Provide EE solutions
- Analyze financial feasibility of EE solutions
- Plan, deliver, install, test, and commission EE solution
- Set up and manage EnMS
- Maintain and sustain EE solutions
- Train others in EE and EnMS

An organization could employ a person with the above-mentioned skill sets as a permanent internal resource, or it could be acquired as an outsourced service (Thollander, Sa, Paramonova, & Cagno, 2015).

Grid Edge Actionable Framework. The grid edge actionable framework (See Figure 12) that employs four main principles for new technologies to be adapted to the electrical grids. The four principles are listed below.

Principle 1: Redesign regulatory paradigm which involves changing, advancing, and reforming regulation to play new roles in how distribution networks are fully integrated in terms of innovation and operations.

Principle 2: Deploy enabling infrastructure involves having the proper infrastructure to support a new business model and the future energy system in a timely manner.

Principle 3: Refined customer experience is to have customer engagement by making the experience easier, convenient, and economical.

Principle 4: To embrace a new business model is to pursue a new revenue source and to develop business models to adapt to the Fourth Industrial Revolution (World Economic Forum, 2017)

The grid edge actionable framework can be adapted to be drivers to overcome barriers identified in this research. Principle 1 can be utilized for Group 1 barriers, Principle 2 for Group 2 and 3 barriers, Principle 3 for Group 3 barriers, and lastly, Principle 4 for Group 2 barriers.

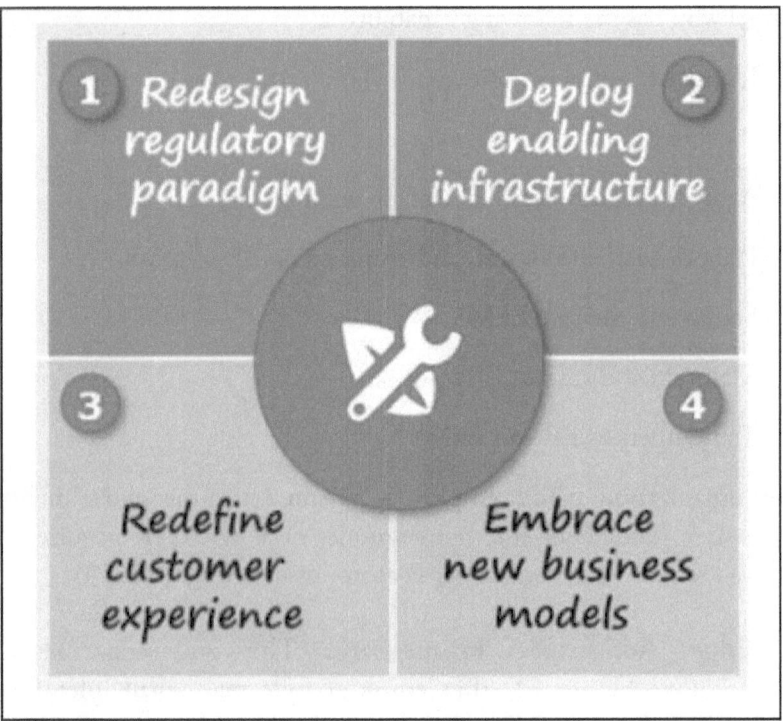

Figure 12 Grid Edge Actionable Framework. Adopted from "The Future of Electricity New Technologies Transforming the Grid Edge", World Economic Forum (2017).

The key components missing from the current grid edge actionable framework to meet the research gaps of this study are defining opportunity and defining implementation procedures and methods. The first component of this framework emphasizes the redesign of regulatory paradigm. If EE opportunities cannot be identified, then carrying out the redesign of a regulatory paradigm will have no effect on the overall outcome of an EE initiative (Kadam, 2014). Component 4 of this framework relates to embracing new business models, and in the context of EE, it means finding new EE solutions. Even if new EE solutions were found, an organization should have prior established implementation procedures and methods before implementing new EE solutions. Therefore, it is essential to have a proven EnMS system in place before applying this framework in an organization (Kelley, Goldberg, Magdon-Ismail, Mertsalov, & Wallace, 2011).

SUMMARY

Chapter two presented a literature review which analyzed the information from previous papers in relation to this study. Various taxonomies of barriers identified by other research were discussed in the adaptation of EE technologies. A literature review was conducted on the recommendation for the implementation of EE technologies such as existing models in EE, use of energy control systems, standardizing energy management system, and various established reports on successful EE implementation. A literature review was carried out on previous research related to the usefulness of EE technologies in areas of market gains and economic development, and lastly, a literature review was done on existing frameworks on competency and behavioral model that could be adapted as the framework for this research.

Chapter three sets out the selected research design of this study and describes the procedures of the inquiry or strategies that were followed. Chapter three also describes the specific methods of data collection, analysis, and interpretations employed. Chapter four sets up the analysis and presents the results of this study. Finally, chapter five encompasses the summary, conclusions, and implications of this research.

Chapter 3
Research Methodology

The present chapter describes the type of research methodology adopted in the present study. Research methodology is a process of evaluating and the assessing of information from data collected to achieve the research objective (Hassani, 2017). In this chapter, the researcher elaborates on the type of research methodology adopted, the data collection methods used, the type of sampling performed, and the justifications for the selection of the specific research methodology. Furthermore, the utility and credibility of the research instruments used are also described (Bolarinwa, 2015).

RESEARCH DESIGN

From all of the possible approaches that are generally found, specific areas that can be used for behavioral analysis are identified. There has to be a certain level of knowledge previously gained in order to understand the subject; hence, the researcher's prior knowledge on this subject will be used to conduct the research (Saunders, Lewis, & Thornhill, 2012). First, the most important part of the research methodology followed by Saunders, Lewis, and Thornhill (2012) is the choice between explanatory, exploratory, and descriptive research design. Exploratory research design helps in the understanding and identification of certain phenomenon within the classification of research questions in order to have a clear concept for the result of this research (Saunders, Lewis, & Thornhill, 2012). Descriptive research design is used to find out about any particular issue or problem that is related to an event and any causes that relate to the characteristics of a problem (Collis & Hussey, 2013). Based on these two designs, the choice is either exploratory or descriptive. Explanatory research design helps in the identification of any relationship or association between any variables that are related to the acts and happenings around any issues or a fact (Saunders, Lewis, & Thornhill, 2012).

Since this research aims to understand the perceptions of energy experts toward commercial and industrial electricity consumers on their adoption of EE technologies, the phenomenological research design was chosen. It justifies and provides or identifies any behavioral pattern in the adaptation of EE technologies from all of the available strategies. An exploration to gather information is done in order to understand the area of research and the specific phenomenon that was previously not examined in the study region (Anon, 2015). Among all the regions, the Klang Valley, Malaysia was chosen since it is the source for mass consumption of EE technologies compared to any other country with the same demographics. It is revealed that Malaysia focuses a large amount of trade and investment in manufacturing sectors, and it was found that there is a reduced amount of awareness or measures taken to adopt EE technologies, apart from the region being the researcher's hometown (Sadler, 2017).

A research design involves determining the appropriate research philosophy, research methodology (approaches), data collection strategies, data choices and technique, and procedures for data collections (Saunders, Lewis, & Thornhill, 2012). Saunders, Lewis, and Thornhill (2012) designed a model called "The Research Onion" (see Figure 13) that is used as a layer after a layer method to carry out a research design and to address the objectives' actions. The Research Onion framework is where the research design is decided from the outmost layer all the way to the innermost layer similar to peeling a real onion. It has five layers, and each layer will have some type of possible research decision to be made appropriate to the research study.

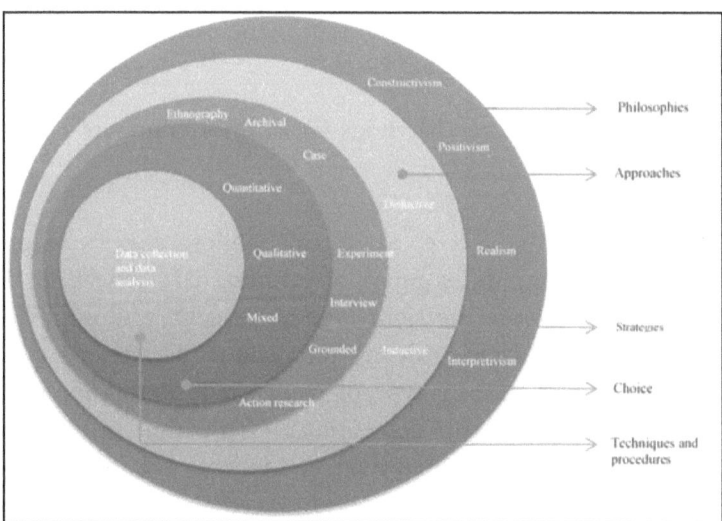

Figure 13 Saunders's Research Onion Methodology. Adapted from "Research Methods for Business Students", by Saunders, Lewis, and Thornhill (2012).

Table 5 Research design utilizing Saunders's "Research Onion Methodology"

Research Design Elements	Proposed Method	Chosen Method
Research Philosophy	Ontology, Epistemology, Ethics, Logic, Phenomenology	Phenomenology
Research Philosophy Stances	Positivism, Interpretivism, Constructivism, and Realism	Interpretivism
Research Approaches	Deductive, Inductive, and Mixed	Inductive
Research Strategies	Experiment, Survey, Case study, Grounded theory, Ethnography, Action research, Qualitative interview	Qualitative interview
Research Choices	Qualitative, Quantitative, and Mixed	Qualitative
Research Techniques and Procedures	Sampling, Secondary data, Observation, Interviews, Questionnaires	Interviews (primary data) and Secondary data

Note. Adapted from "Research Methods for Business Students", by Saunders et al., 2012.

Research Philosophy. There are many philosophies related to the identification, collection, and support of primary data collection. The aim of this section is to choose the right type of research philosophy in order for the researcher to have a guideline for conducting the research. These philosophies shed light on any type of choice related to the examination of research from any of the perceptions. With this understanding, a selection procedure will be identified (Saunders, Lewis, & Thornhill, 2012). Traditionally, research philosophy includes at least five core fields or disciplines. These fields are ontology, epistemology, ethics, and phenomenology (Smith, 2016). All of these are excluded from the research onion diagram (see Figure 13).

Ontology. Bryman's (2012) concepts found that ontology is about the way an individual perceives based on how he or she interprets or understand the question based on his or her social system of belief, e.g. study of being or what is. How the individual perceives is related to the activities of social entities, which depend on the perceptions of social actors. There are two types of possible research paradigms related to the social sciences resulting in either interpretivism or positivism (Ling & Ling, 2015). The perceptions of the energy experts will directly relate to external and internal factors that influence the CIEC on the adaptation of EE technologies (Bryman & Bell, 2015; Collis & Hussey, 2013). Social phenomenon relates to the perceptions of constructivism from the perceptions of a social phenomenon that will constantly change since it is a behavioral change (Bryman & Bell, 2015).

From the viewpoint of ontology, the natural direction would be to identify behavioral patterns related to the research problem. With many strategies that were proposed in relation to the adaptation, there should be some identical part of an approach that is normally followed. With this particular type of research, there should be an ontological approach since it is taken from the participant's perspective (Saunders, Lewis, & Thornhill, 2012). Constructivism is a type of approach relating to interpretivism. This approach will help in understanding the behaviors related to the perspectives of participants and their understanding of adaptability issues related to the research problem (Thanh & Thanh, 2015). Ontology concentrates on what is the individual's perception of a social or system of belief. The present research aims to examine the resistance of electricity consumers to the adoption of energy-efficient postures.

Epistemology. Epistemology is the study of knowledge and will help in answering queries related to issues that reveal the individual's knowledge, for example, the study of knowledge. Many assumptions related to social science state that basic knowledge is required to understand any concept. Epistemology offers two different philosophical ideas of how an individual knows: the positivist point of view and interpretivism (Saunders, Lewis, & Thornhill, 2012). Positivist theories hold that universally accepted laws and beliefs that are based on the subject matter of interpretations from a quantitative approach attempt to find the association from the controller and the predictor (O'Brien & Scot, 2012). The approach of an interpretivist is to use qualitative methodology to interpret an individual's perception on a particular topic from his or her personal experience in order to gain knowledge (Bryman, 2012) as in this research the personal experience of energy experts on CIEC's resistance to the adoption of EE technologies.

Ethics. Ethics is the study of right and wrong actions, is the branch of philosophy that explores the nature of moral virtue, and evaluates human actions (Smith, 2016). Philosophical ethics differs from legal, religious, cultural, and personal approaches to ethics by seeking to conduct the study of morality through a rational, secular outlook grounded in notions of human happiness or well-being (White, 1993). A major advantage of a philosophical approach to ethics is that it avoids the authoritarian basis of law and religion as well as the subjectivity, arbitrariness, and irrationality that may characterize cultural or totally personal moral views (Smith, 2016). Generally speaking, there are two traditions in western philosophical ethics regarding how to determine the ethical character of actions. The first argues that actions have no intrinsic ethical character but acquire their moral status from the consequences that flow from them. The other tradition claims that actions are inherently right or wrong. The former is called a teleological approach to ethical evaluation and the latter is called deontology (Tiesen, 2011; & Smith, 2016). The energy efficiency technology is considered as the right solution to reduce the burden of environmental issues. Furthermore, it is suggested as the best solution to address the demand for energy. The ethical philosophy provides guidelines on how to act in an uncertain situation. These guidelines sometimes do and sometimes do not overlap with the law. People are often times not aware of the consequences and issues regarding environmental problems; hence, this research will make a significant contribution by providing important information to the population.

Phenomenology. Phenomenology is the study of our experiences and structures of consciousness as experienced from the first-person point of view. The central structure of an experience is its intentionality as it is being directed toward an object by virtue of its content or meaning (which represents the object) together with appropriate enabling conditions (Lapan, Quartaroli, & Riemer, 2012).

Phenomenology is literally the study of phenomena, the appearance of things, of things as they appear in our experience, or the different ways we experience objects. In short, phenomenology is about the meanings things have in our experience. Basically, phenomenology studies the structure of various types of experiences ranging from perception, thought, memory, imagination, emotion, desire, and volition to bodily awareness, embodied action, and social activity, including linguistic activity. The structure of these forms of experiences typically involve intentionality, that is, the directness of experience toward things in the world or the property of consciousness that it is a consciousness of or about something (Smith, 2016).

The basic intentional structure of consciousness that we find in reflection or analysis involves further forms of experience. Thus, phenomenology develops a complex account of temporal awareness (within the stream of consciousness), spatial awareness (notably perception), attention (distinguishing focal, marginal, or horizontal awareness), awareness of one's own experiences (self-consciousness in one sense), self-awareness (awareness of oneself), the self in different roles (as thinking, acting, etc.), embodied action (including kinesthetic awareness of one's movement), purpose or intention in action (more or less explicit), awareness of the other person (in empathy, inter-subjectivity, or collectivity), linguistic activity (involving meaning, communication, and understanding others), social interaction (including collective action), and everyday activity in our surrounding life-world (in a particular culture) (Smith, 2016).

Based on the definition of the four fields of philosophy mentioned above, phenomenology research philosophy is the most suitable for this study as it will provide ways to identify the perceptions (spatial awareness) of participants, the lived experience of the energy experts, and the motivation level of CIECs of Klang Valley, Malaysia as they relate to the adoption of EE technology. The reason for focusing on phenomenology is because it describes an event, activity, or phenomenon, which is an appropriate qualitative method. In a phenomenological study, the researcher can use a combination of methods, such as conducting interviews and reading documents. Unlike other qualitative methods, phenomenology does not start with a well-formed hypothesis. Phenomenological studies often conduct a lot of interviews, usually between five and 25 common themes, to build a sufficient dataset to look for merging themes and to use other participants to validate the findings.

Research Philosophy Stance. The philosophical stances associated with philosophies require careful thought as they provide structure, guidance, and possible limitations to the way the researcher can collect data to create valid findings (Matthews & Ross, 2010). Based on Saunders's "Research Onion Methodology", the general philosophical stances are as follows:

Positivism. Positivism generates hypotheses (or research questions) that can be tested and allows explanations that are measured against accepted knowledge of the world we live in. This position creates a body of research that can be replicated by other researchers to generate the same results. The emphasis is on quantifiable results that lend themselves to statistical analysis (Matthews & Ross, 2010). The present research examines the perception of energy experts on CIEC of Klang Valley toward the adoption of EE technology. It is not quantifiable. Positivist philosophy is appropriate to replicate the previous results and theories.

Interpretivism. Interpretivism refers to approaches emphasizing the meaningful nature of people's participation in social and cultural life. Researchers working within this tradition analyze the meanings people confer upon their own and others' actions and take the view that cultural existence and change can be understood by studying what people think about, their ideas, and the meanings that are important to them (May 2011). This approach is more appropriate for the present study because it helps in identifying the reasons for the resistance in the adoption of EE postures.

Constructivism. Constructivism argues the opposite of objectivism. It is a viewpoint that argues that social phenomena are actually constructed by social actors. In research using a constructivist philosophy stance, it is believed that individuals seek understanding of the world they live in and work. Individuals develop subjective meanings of their experiences and meanings directed at certain objects or things. These meanings are varied and multiplied, leading researchers to look for the complexity of views rather than narrowing meaning into a few categories or ideas. The goal of the research relies as much as possible on the participant's views of the situation being studied. The questions become broad and general so that the participants can construct the meaning of a situation, typically forged on discussions or interactions with other persons.

The more open-ended the questioning, the better as the researcher listens carefully to what people say or do in their setting. Often these subjective meanings are negotiated socially and historically. They are not simply imprinted on individuals but are formed through interaction with others and through historical and cultural norms that operate in individual lives. Thus, constructivist researchers often address the process of intersection among individuals. They also focus on the specific contexts in which people live and work in order to understand the historical and cultural setting of the participants. Researchers recognize that their own backgrounds shape their interpretation, and they position themselves in the research to acknowledge how their interpretations flow from their personal, cultural, and historical experiences. The researcher's intent is to make sure or interpret the meanings others have about the world. Rather than starting with a theory, inquirers generate or inductively develop a theory or pattern of meaning (Matthews & Ross, 2010).

Realism. Realism is similar to positivism in its processes and belief that social reality and the researcher are independent of each other, for example, the participants and others will not create biased results. However, where the two theories differ is in realism being grounded in the idea that scientific methods are not perfect. The theory believes that all theories can be revised, and therefore, scientific methods are not perfect. It believes that all theories can be revised and that our ability to know for certain what reality is may not exist without continually researching and leaving our minds open to using new methods of research (Lapan, Quartaroli, & Riemer, 2012).

Based on the definitions of the above-mentioned general philosophical stances, this research chooses interpretivism as the research philosophical stance for the present study (Saunders, Lewis, & Thornhill, 2012). An interpretivist philosophy is best suited for the present research project due to the following specific reasons. First, the researcher attempted to understand the perceptions of energy consumers that cannot be quantified but could be qualified. Second, the use of numerical data methods was deemed not useful to understand a specific research phenomenon since quantifiable assessments do not provide in-depth insights. By being an interpretivist, the researcher was able to understand the in-depth insights of CIEC's resistance to the adoption of EE technologies through the perception of the energy experts and was able to arrive at a better consensus regarding the situation. Hence, an interpretivist philosophy stance was adopted.

Research Approaches. The selection of research approaches depends strongly on the decision in selecting the philosophical stance as mentioned in the previous section with the need to assess the research aims, limitations associated with the study, and the researcher's personal opinion to decide which will work best for the work. There are three types of research approaches: deductive, inductive, and mixed (deductive plus inductive).

Deductive. Deductive means that the research starts with a statement or question and the research sets out to answer it (see Figure 14). The aim would be to conclude with a "yes" or a "no" response to the question. Questions may be statements or informed speculation about the topic that the researcher believes can be answered. The process of deduction moves from theory to the research question, to data collection, findings, and then finally to rejection or confirmation of the research question. This should lead to a revision of the theory and often starts the process over again (Dudovskiy, 2015).

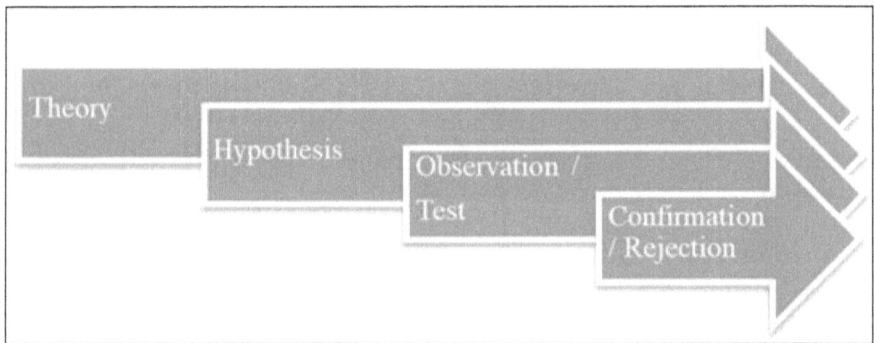

Figure 14 Deductive process. Adapted from "The Ultimate Guide to Writing A Dissertation in Business Studies – A Step-by-Step Assistance", by J. Dudovskiy, 2015, p. 28. Copyright@2015 by research-methodology.net.

Inductive. An inductive approach to research entails a process of creating a theory (see Figure 15). The process moves in the opposite direction of the deduction approach, taking focus from the working title of the researcher and not the existing theory. This means the research goes from research question to observation and description to analysis and finally to theory. Therefore, if little research exists on a topic, then an inductive approach may be the best way to proceed (Dudovskiy, 2015).

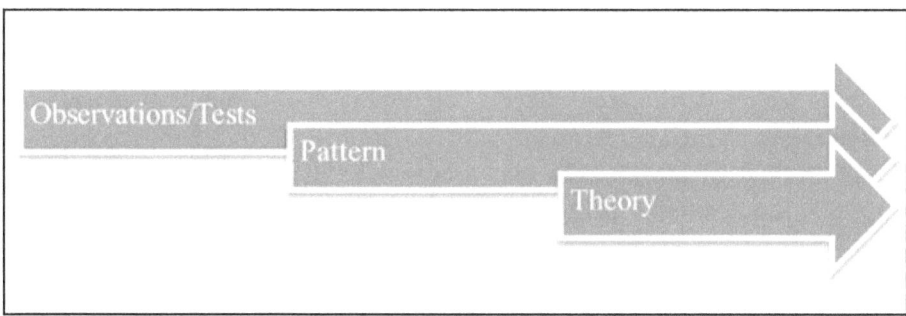

Figure 15 Inductive Process. Adapted from "The Ultimate Guide to Writing A Dissertation in Business Studies – A Step-by-Step Assistance", by J. Dudovskiy, 2015, p. 28. Copyright@2015 by research-methodology.net.

Mixed. A mixed research method is basically a combination of deductive and inductive approaches applied so that both quantitative and qualitative findings are founded in the study. The findings of the study are presented both in numbers, for example, quantity and measurements, and with data such as personal accounts, opinions, and description (Creswell, 2014).

The present research aims to present the possibility of energy-efficient technology being implemented in Klang Valley, Malaysia. Therefore, an inductive approach is appropriate for the present research. By the end of this research, specific information is compiled that addresses the research objectives as a whole. This research chose inductive as the research approach since the present study goes from research questions to description to analysis and, finally, theory. An inductive research approach is best to create a theory about a phenomenon that does not exist prior in the literature. Being a novel research phenomenon, the researcher chose the inductive approach.

Research Strategy. Research strategy will give an idea on how the research can be conducted with respect to addressing the research questions (Saunders, Lewis, & Thornhill, 2012). There are several classifications of research strategies that are followed such as action research, grounded theory, interviews, survey, case study, archival research, and ethnography (Saunders, Lewis, & Thornhill, 2012). The present study looks for comprehensive details about the phenomenon that is related in interpreting behaviors of people who are involved in the study, and based on this reasoning, the utilization of qualitative interview was chosen.

Research Choice. Most of the studies that are related to the identification of any behavior follow a qualitative perception. Previous research relates to the relativity and adaptability of a qualitative research methodology (Gareis, Heumann, & Martinuzzi, 2009; Labuschagne & Brent, 2006; Michaelides, Bryde, & Ohaeri, 2014; Silvius & Schipper, 2009; Taylor, 2008; Turner, 2010; Wang, 2014). Hence, this research methodology used findings to interpret behavioral concepts. Interpretivism was fit into the concept of epistemological perceptions that related to understanding the viewpoint of the participants. From the researches of Antwi & Hamza (2015) and Lincoln & Guba (1985), rather than a generalized term from a quantitative perception, this research followed the results that gave specific solutions or findings (Antwi & Hamza, 2015; Lincoln & Guba, 1985). From the above criteria, a qualitative approach was taken with a descriptive and interpretive approach that confided with the present study. It will help in exploring the characteristics and experiences of the participants, that is, energy experts, or any of the experiences that are related to the research topic.

Research Techniques and Procedures. Within many parts of this research, data collection was the most crucial part of this study. According to Merriam and Tisdell (2016), qualitative research is a type of method involving different peoples' perspectives toward the adaptation of EE technologies (Merriam & Tisdell, 2016). Any type of qualitative research method will include non-numerical data that relates to the participants' (energy experts) experiences and knowledge (Saunders, Lewis, & Thornhill, 2012).

This research clearly approached the social entities that are related to constructivism. It provides evidence that barriers are inherent in the development stages of a product because of these factors. This study fills the gap that is identified from the data that will be collected from qualitative interviews. This study considers the integration of various environmental dimensions from the field of electrical consumers of the commercial and industrial sectors (CIECs). The data collected identifies the many gaps that are produced affecting basic knowledge or knowledge that already exists. This would be a natural approach to a phenomenological research study. With many ongoing studies that are been conducted, this study will support the effectiveness in the adaptation of sustainable implementation of EE technologies. The guidelines that are proposed by several different authors indicate that any terms of failure while addressing the issue will lead to sustainability issues. However, there are many gaps that have to be identified with the adaptability of EE technologies, and there should be a proper procedure in place when attempting to introduce such technologies.

Selection Criteria. This research concentrates only on Klang Valley, Malaysia area since this region is found to be using more electrical energy than any other identical demographical county that has commercial and industrial sectors. There are many complexities in the commercial and industrial sectors when implementing EE technologies. Data collection will be done with participants in the Klang Valley, Malaysia region since the data collected is reflective of a typical metropolis region for any country in the Association of Southeast Asian Nations (ASEAN). The region of Klang Valley, Malaysia is a leading subject in the availability of any trade or educational factors. It is considered one of the strategical regions related to the implementation of various kinds of EE technologies (APEC, 2011). In this region, there are many identifiable perceptions of energy experts on commercial and industrial electricity consumers toward the adoption of such technologies.

Selection of Participants. Among the commercial and industrial sectors of Klang Valley, Malaysia, there are different types of classifications based on their business and size. In this study, the participants selected are not based on any particular business or size. With the inductive approach, there are many divisions of a sector, all of which have a high standard of availability for data collection, universal size, availability of any such resource, and the nature of the study (Wright, Klein, & Wieczorek, 2015). According to Lobe, Livingstone, and Simões (2008), a chosen research methodology is based on the objectives that have been framed for such research in carrying out the data collection procedure. Qualitative research methodology is used for this research since the major objective of this study revolves around the effect of the behavioral factor that is related to the research objectives. With energy consumers being the sample for the research and participants for data collection, it helps in the selection process to confirm that the sampling method is valid since it relates only to the adaptation of behaviors.

RESEARCH QUESTIONS

This study will address the following research questions:

RQ #1. Why are electricity consumers in the Klang Valley resistant to adopting energy-efficient postures?

RQ #2. How can energy-efficient technologies be implemented in Klang Valley, Malaysia?

RQ #3. How do consumers find energy-efficient technologies useful?

POPULATION AND SAMPLING STRATEGY

In research, sampling is the process of selecting participants who will take part in a study on the basis that they provide responses relevant to the context of the research. For a qualitative research project, researchers recommend the use of judgment or purposive sampling and convenience sampling. Convenience sampling implies the selection of participants for the research based on the convenience of the researcher; the sampling technique demands less cost and time, but there are chances of acquiring poor quality data. In purposive sampling, the researcher selects participants according to the needs of the study, with participants having a broad general knowledge of the research topic or those who have undergone the experiences and whose experience is relevant to the research topic. The reason for selecting this category of participants is to allow the entire range of experience and the breadth of the concept or phenomena to be understood and to create a range of variation in the phenomenon of the research study (Oppong, 2013).

Sampling is generally conducted on the basis of access, representativeness, and strategy (Bryman, 2012). For the present study, both purposive and convenience sampling could be used. This sampling technique enables cross-checking the eligibility of the participants on the basis of the research objectives (Easterby-Smith, Lewis, &, Thornhill, 2012). In this regard, samples selected were based on the convenience of the researcher and CIEC are selected purposefully, which is necessitated by the research.

The intended strategy for data collection is represented by the experiences of energy experts who address the basis of strategy, access, size, and representativeness (Bryman A., 2012). Only knowledgeable participants in energy efficiency implementation have the ability to answer the open-ended, semi-structured questions. The sample size for such a procedure is purely based on face-to-face intentions and the ability in answering such questions (Bryman A. , 2012). Since this is a qualitative study, the answer given by the experts should be beneficial, and they can be answered by any one of the issues, instead of a generalized result. This methodology type is identical to the sample size that is prescribed to get more comprehensive knowledge on this data (Silverman, 2015). The present study involves 10 participants for the qualitative data collection. Participants with experiences on the implementation of EE technologies are most likely to be eligible. Participants will confide in the researcher, and the information provided by the partakers will relate to the implementation of EE technologies.

RESEARCH INSTRUMENT

The researcher is the primary instrument or medium through which the research was conducted. The quality of observation is contingent on the expertise of the researcher who serves as an instrument in generating data. The researcher is the research instrument and has the ability to observe "mundane" details, to conduct in-depth interviews, and to reflect on the meaning of observation and interview data which are essential to the success of the research (Xu & Storr, 2012). The following subsections describe the instruments utilized by the researcher to ensure the utility and credibility of this research study.

Utility and Credibility of the Research. The term utility of qualitative research depends on 'how users perceive the different characteristics of the study such as reliability, validity, and credibility of the researcher' (Sun, 2009). The terms credibility and utility are directly related to trustworthiness and examine credibility, transferability, dependability, and conformability of the research (Hannes, 2011). To validate the trustworthiness of this research, the following steps were followed. First, the interview guide was prepared based on the examination of previous research (literature review) conducted in the same research context (Gareis, Heumann, & Martinuzzi, 2009; Turner, 2010; Silvius, Brink, & Köhler, 2010). Second, a pilot study was conducted with one to two participants (Quinlan, Zikmund, Babin, Carr, & Griffin, 2015).

To improve the credibility of the research study, verbatim quoted transcripts of the participants during the interview were sent to the participants to check and confirm that the facts and information they provided were accurate, clear, and reflect their views by the way of member checking wherein the themes of the interview were summarized (Harvey, 2015). Appendix E contains a copy of the "Interviewee's Transcript Review (ITR)" letter requesting the research participant to check and confirm that the facts and information in their interview transcript are accurate, clear, and reflect of their views.

Role of the Researcher. The role of the researcher is imperative in any study. The researcher has to determine what, why, and how the research study is to be conducted, and upon determining these parameters, he or she has to plan and prepare methodological procedures for the research, including the preparation of a consent form to participate in the research. The researcher had to design the interview guide wherein the interview structure was determined. The researcher collected all secondary sources pertinent to the research context where existing theories and models (literature review) were examined. Second, the researcher conducted interviews with respect to the research topic. Participants were then contacted through e-mail addresses retrieved from online sources. The researcher further transcribed raw data to a digitized format and examined the collected data thematically. The researcher defined and named the categories of data, e.g. starting with identifying the codes for the data. The researcher further determined the elements where the codes would be linked by utilizing an analysis technique where the codes came together in one overall analysis to provide a conceptualized 'web on meanings' based on the themes and sub-themes derived. The data was presented in this sequence. Finally, the researcher utilized his personal knowledge and experiences to make sense of the data presented, summarized the data, and presents it in the Discussion and Conclusion section highlighting major findings. During the entire process of the research, the researcher maintained a non-bias stance (Roller & Lavrakas, 2015).

Researcher Bias. In order to conduct the interviews in a non-biased manner, the procedures for data collection were conducted in two phases, preparation of the interview question and the preparation for conducting the interviews.

In preparing the interview questions, the researcher ensured that the following types of bias questions were avoided to ensure the accuracy of data collected:

- Leading questions bias, where the questions suggest what the answer should be.

- Misunderstood question bias, where words, context, culture, and different interpretations of words and sentences cause misunderstanding. The questions were designed as simple and clear.

- Unanswerable question bias, where participants are not able to answer due to the lack of knowledge on a particular question.

- Question order bias, in which the order in which the questions are asked can lead to a biased answer.

When conducting the interview, the researcher observed the following to ensure the accuracy of data:

- Will always be neutral in his or her thoughts and should not be influenced by one's previous knowledge or experience on the research topic.

- Will always control his or her expressions, body language, tone, manner of dress, and style of language that might introduce bias and should remain as neutral in one's dressing, tone, and body language during the course of the interviews.

- Will not give opinions while conducting the interview.

- Will avoid asking bias questions which may lead the participants.

- Ensure that each time before conducting the interview, the researcher will go through one's interview guide so that he or she is always reminded of the things to avoid while conducting the interview (Roulston & Shelton, 2015).

Sources of Data (triangulation). In qualitative research, dependability addresses whether the research process is clearly documented, logical, and traceable, specifically on the chosen methods and the decision of the researcher (Hannes, 2011).

One method to ensure the dependability of a research is to employ triangulation. Triangulation is generally used to strengthen the dependability of the research by checking if inferences drawn from one set of data source compare with the data from the other source of data (Creswell, 2014). For this research, the type of triangulation method utilized is data source triangulation, which involves the comparison of data that relates to the same phenomenon but from different types of data sources consisting of interviews, literatures, and governmental reports. Generally, there are two types of data collection that are generally carried out, namely primary and secondary data (Creswell, 2014).

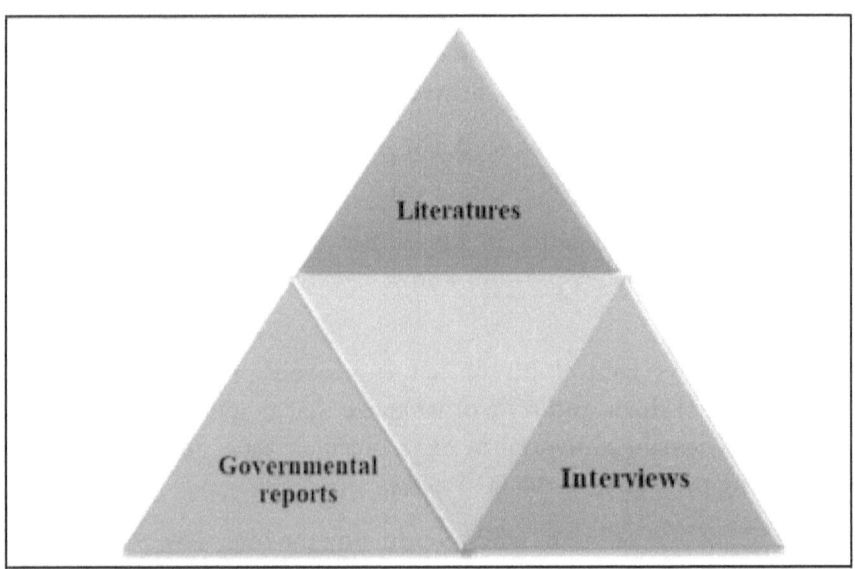

Figure 16 Data Source Triangulation. Adopted from "Peeling Saunders's Research Onion", by Sahay, A.A. (2016).

As described in Figure 16, for this research the primary source of data came from interviews, and the secondary source of data came from literature review and collection of governmental energy reports. It was found that the secondary reports that are related to the annual report identified information that was related to the commercial and industrial sectors.

DATA COLLECTION PROCEDURES

Data Collection Methods. Most of the primary data for this research that was collected came from interviews. The type of data collection method is classified into three types of structures: semi-structured, unstructured, and predefined structures. First, the data has to be classified with primary data and then by the secondary data collection (Creswell, 2014). The interview method that is classified will use the information that is found within the collection of primary data whereas semi-structured and unstructured interviews will not have a predefined structure; hence, it will have open-ended questions (Creswell, 2014; Denzin & Lincoln, 2017; Myers & Newman, 2007; Yin, 2009). This research followed the semi-structured interview technique. This technique helped in exploring unique experiences when creating a particular project with leaders (Creswell, 2014). The technique enabled the participant to focus on the discussion that was within the investigation (Easterby-Smith, Lewis, & Thornhill, 2012). The interview questions were based on two sets of questions that related to the semi-structured interview process with the preparation of a detailed literature structure. From the findings of certain behavioral factors, there can be a certain amount of guided procedures that are related to the initiatives from the top management of commercial and industrial consumers of electricity where the participant is an energy expert providing services to CIECs. The interview questions can be classified according to the participant's roles and responsibility related to the implementation of EE technologies. With a properly guided interview, participants are able to express their diverse views on the research topic.

The questions asked in the face-to-face, opened-ended, semi-structured interview will provide the participants the freedom of expression in answering the questions in such a way that it will increase the level of their interest in the research topic. This provides valuable insight to the study. This research will be useful as a guide to promote adoptability or motivation with sustainability with respect to the present study. With the reduction of barriers related to the adaptation of EE technology, it will give a perspective in deriving useful drivers to overcome the lack of implementation of these technologies. For this study, an interactive approach was adopted for data collection (Denzin & Lincoln, 2017). The interactive approach allows for the researcher to successfully identify solutions from the participant's point of view.

Selection of Interview Questions. The selection of interview questions for this study was based on the examination of previous research conducted (literature review) on the context of efficient energy management. All researches conducted on this topic in developed and developing countries were assessed. The interview questions were designed based on the research questions, which aimed at exploring the perceptions of CIEC in Klang Valley, Malaysia. Once the questions were designed, an approval from the supervisor (a faculty member assigned for this research upon the approval by the SMC Academic Research Board) was acquired. This was not a single step process. Personal bias of the researcher was reduced after discussions with the supervisor, who aided in enhancing the questions. A copy of the interview protocol used for the interview is provided in Appendix D.

Conducting the Study. For this research, face-to-face interviews were conducted after prior appointments were made with participants in Klang Valley, Malaysia. To contact the participants, e-mail addresses were retrieved from the internet for energy experts serving CIEC in Klang Valley, Malaysia. The Consent Form (see Appendix A) to participate in this research was mailed to the selected participants. Upon the participants' consent to participate in the research, an appointment for the interview was confirmed through emails. The duration and time for the interview were fixed with the participants of the research, and their convenience was also taken into consideration. Since the research attempts to examine the perceptions of CIEC, the selected participants were approached at their offices based on their schedule. Participants were provided an option to select any of two languages: English or Malay. The average time for the interview with a participant was around 45–60 minutes. Audio recording and notes were used to collect raw data from the participants by observing participants' behavior and from the face-to-face interviews. After the collection of data in the form of audio files and notes, the data was transcribed to digital form where the actual verbatim quotes of the participants were captured. All data were converted to a word document in the English language on the same day the face-to-face interview was conducted in order to reduce the chances of lost data.

Secondary Data Collection. Secondary data collection conducted in relation to the party related to this research topic may be combined with other data collected. If any researchers use these data, it will satisfy with the necessity based on a different timeline from the past. The documents that are written based on any type of written procedures will be satisfied in electronic form (Sindhu, 2012). With the variety of information that is related in the collection of data, the focus of the data was toward EE technology, which has demands in the marketplace. Secondary data that is crossed over has its implication within the necessity of the research problem. This secondary data was classified either with internal or external factors. Here the internal factor implies that comprehensive information was featured within CIEC of Klang Valley, Malaysia. External factors were sources that related to various external sources outside of CIEC of Klang Valley, Malaysia (Sindhu, 2012).

DATA ANALYSES

One of the central issues in qualitative research is the way through which responses of the participants with social reality and subjective meanings are conveyed in the report (Hesse-Biber, 2011). Data collected using qualitative data collection instruments do not have simple meanings and involve several interpretations that need to be appropriately assessed. Researchers should incorporate efforts to identify such meanings and these should then be conveyed in the form of themes (Bazaley, 2009; Lapan, Quartaroli, & Riemer, 2012).

In this research, thematic analysis was conducted on the basis of the following steps. First, a code was developed either in a word or short phrase that represents a theme or an idea. It was then assigned a meaningful title, which were the elements that were coded. There are three types of coding, open, axial, and selective coding. The present study adopted axial coding, where the core themes were disaggregated during the qualitative data analysis. Second, textual data was categorized by identifying common patterns within the responses to the themes and sub-themes of the research, where the critical thinking skills of the researcher played a critical role in data analysis. For the analysis of textual data, the following interpretation method was used:

- Word and phrase repetitions
- Data triangulation
- Search for missing information
- Metaphors and analogs

The above-mentioned interpretation method was completed with the aid of Computer Assisted Qualitative Data Analysis Software (CAQDAS) such as QSR Nvivo software, which is a qualitative analysis software used to analyze unstructured or non-numerical data. The software was used to analyze textual data based on commonly occurring words.

Lastly, the analyzed data was summarized by linking the findings to the objectives of research and the interrelationship of themes and sub-themes, where noteworthy quotations from the transcripts and major themes and possible contradictions within the findings were highlighted.

In order to enhance utility and credibility, there have to be several analytical strategies that were found to be feasible for applying (Leung, 2015; Lincoln & Guba, 1985). The questions were carefully evaluated with help from carefully constructed interviews. The interviews that were conducted had a relation to the effects of behavioral analysis among participants. After one set of interviews, there has to be a perfect evaluation or follow up of the questions that will be conducted. Later, these were member-checked with the interview candidates, resulting in removing irrelevant or redundant type of questions which were entered into the data collection procedure. The coding procedure that has been indicated had no effect over any type of codes related to the procedures.

ETHICAL CONSIDERATIONS

The interviewer was aware of the potential conflicts of interest, which could have occurred when the discussion infringed on sensitive or personal information. Upon first contact with the interviewees, the interviewer provided them with a letter from the Swiss Management Centre (SMC) University validating that the interviewer was conducting important research within the scope of important educational goals. Since the interviewer was aware of the privacy and confidentiality of all participants, necessary steps were taken to protect their privacy. All selected energy experts who are involved with Commercial and Industrial Electricity Consumers (CIEC) were not forced or compelled to participate in this research. After their approval, an invitation from the interviewer was sent through an e-mail or via phone, and a mutually agreed upon time was scheduled for the interview. Anonymity and flexibility regarding their personal information were preserved, and the choice of the individuals to discontinue during the interview was provided. After the completion of this research project, any typed and videotaped interviews will be destroyed within three years after the successful defense of this research dissertation. Interviewees signed an official letter explaining that the collected data would be used only for the dissertation's research. SMC University officials were asked to verify the level of integrity of this research. The published version of this research will be reported anonymously based on Ethical Principles of Psychologists and Code of Conduct, Section 4: Privacy and Confidentiality (APA, 2017).

SCOPE AND DELIMITATION OF THE STUDY

The present research was conducted entirely in Klang Valley, Malaysia, where energy experts serving this region were interviewed to understand the perception of these energy experts on the resistance of CIEC toward the adoption of EE technologies. This study supports the understanding of the behaviors relating to the value created in the field of any sector. In order to conduct the study in the manner described, a semi-structured interview was required. From this, both phenomenon and inductive approaches were identified, and no personal information will be released nor any confiding with respondents.

SUMMARY

Chapter three, on the whole, described the procedures that were taken to identify the behavior of commercial and industrial sector consumers of electricity toward the adoption of EE technologies. This chapter analyzed and selected appropriate elements that best fit the research study. It started with the outermost layer of the research onion framework to its core layer, which is the research philosophies, approach, strategies, choice, and techniques and procedure followed by data collection method, sample size, the types of data collection, and taking into consideration the ethical issues related to participants. Chapter four covers the results of the research, where the results are derived through the methodology defined in chapter three. Finally, chapter five consists of a discussion and conclusion wherein the results of this study are examined to derive what attitudes and behaviors prevent the adaptation of energy-efficient technologies. In addition, the conclusions to the study are derived and recommendations for future research are revealed.

Chapter 4
Analysis and Presentation of Results

The purpose of this qualitative phenomenological study is to investigate and report the lived experiences of respondents participating in this research who are energy experts. This research developed a deeper understanding of the attitudes of commercial and industrial electricity consumers in Klang Valley, Malaysia, and the reasons why CIECs resist the adoption of energy efficiency (EE) postures. Additionally, this research examined the implementation of EE technologies and the benefits of utilizing EE technologies. Understanding lived experiences allows for the phenomenological study of the structures of subjective experience and consciousness as well as a method (Lapan, Quartaroli & Riemer, 2012). In the current study, the participants' responses collected through face-to-face interviews were thematically analyzed.

Chapter four presents the analysis and the results of this research study. This chapter presents the major findings and discussions of this research. It consists of the following four sections: section one presents the demographic statistics of the participants; section two provides details of data analysis; section three presents the results of the study; and section four provides the summary of themes and major findings.

This qualitative research employed a phenomenological approach that provided a specific context in which the researcher explored the practices of energy efficiency (EE) among commercial and industrial electricity consumers within the Klang Valley. Data was collected from the perspective of a particular group of ten participants who were energy experts and were selected from those involved in the Klang Valley's energy services (Creswell, 2014; Saldana & Leavy, 2011; Sun, 2009). The participants selected and interviewed provided real-world perspectives of a complex contemporary phenomenon, namely the impact of EE in Klang Valley.

The Klang Valley has a diverse, energetic, and dynamic population of around 7.2 million people who reside in an area of over 243 km^2 (Department of Statistics Malaysia, 2017). The Klang Valley metropolis region is faced with issues arising from urbanization and has a need to attract more investments. This region accounts for 20 percent of Malaysia's population, and it added RM263 billion to the gross national income in 2010 (Inside Investor, 2012), and its 7.2 million people in this area will affect 40% of the GDP of Malaysia.

DEMOGRAPHIC STATISTICS

The 10 participants in this study were comprised of Malaysian citizens who are qualified energy experts serving the commercial and industrial electricity consumers (CIEC) of Klang Valley. The participants selected for inclusion in this qualitative phenomenology research study are owners of companies providing energy services and those who are employed as directors or energy managers of companies providing energy management services (ESCOs). The selection method included those individuals who could ensure that they were able to contribute to the research from their personal experiences and had an understanding of the research problem (Creswell, 2014). Furthermore, to ensure the participants were able to contribute to the research effectively, the following characteristics were considered: (a) that the participant was experienced and involved in the energy management services for two years or more; (b) that the participant was still active in the energy management services; (c) that the participant has been providing his company's services to the commercial and industrial sector in Klang Valley; (d) that the participant had the knowledge of and has experienced the issues considered under this study; and (e) that the participant was willing to participate in the research study.

To ensure the privacy of all participants was maintained and that they remained anonymous as promised in the signed consent form, code names instead of their actual names were used to reference participants in the interview transcripts and in the representation and presentation of the research. Consequently, to keep code-naming convention simple, the researcher has chosen to refer to all ten participants in this study as PX (an abbreviation where P represents 'Participant' and the letter "X" represents a participant number between one to ten), e.g. P1, P2, P3, P4, P5, etc. Table 6, Figure 17, and Figure 18 provide further demographic statistics of all research participants involved in the study. Table 6 provides participant characteristics relating to their age and gender, Figure 17 provides participant characteristics relating to academic and professional qualifications, and Figure 18 provides participant characteristics relating to their years of experience as energy experts.

All the participants were males who reside in Klang Valley. Their ages ranged from 30 to 60 years old. Tables provide a summary of their age and gender.

Table 6 Participants Characteristics: Age and Gender

Participant Code	Age	Gender
P1	30	Male
P2	32	Male
P3	53	Male
P4	46	Male
P5	53	Male
P6	53	Male
P7	45	Male
P8	56	Male
P9	54	Male
P10	60	Male

Note: Table derived by author from analysis of primary data collection.

Of the 10 participants, four participants were owners of ESCOs, playing key roles in running their business enterprises, four participants held senior managerial appointments responsible for managing business operations and a diverse number of employees, including energy managers, and the rest were employees of ESCOs. All ten participants were qualified energy experts. Nine of the 10 hold Registered Electrical Energy Manager (REEM) certification, seven have bachelor's degrees in engineering, one has a Master of Science degree in energy, one has a Master of Business Administration (MBA), and four have Doctor of Philosophy (Ph.D) degrees in mechanical engineering. Of the 10 participants, six are practicing professional engineers (PEs). The summary of these findings is summarized in Figure 17.

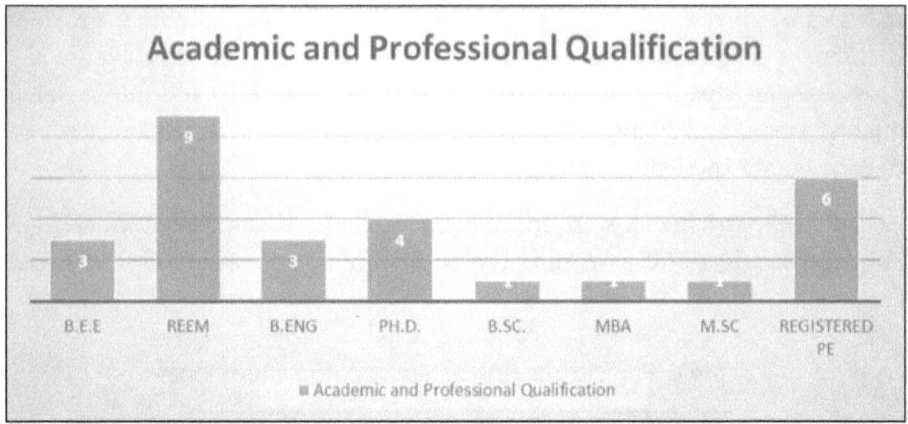

Figure 17 Summary of Academic and Professional Qualifications. Source: Figure derived by author from the analysis of primary data collection.

The participants' experience as energy experts range between three to 30 years; hence, they have an average experience of approximately 11 years. Five participants reported having three to 10 years of experience, and the remaining five claimed to have 11 to 30 years of energy services experience (see Figure 18).

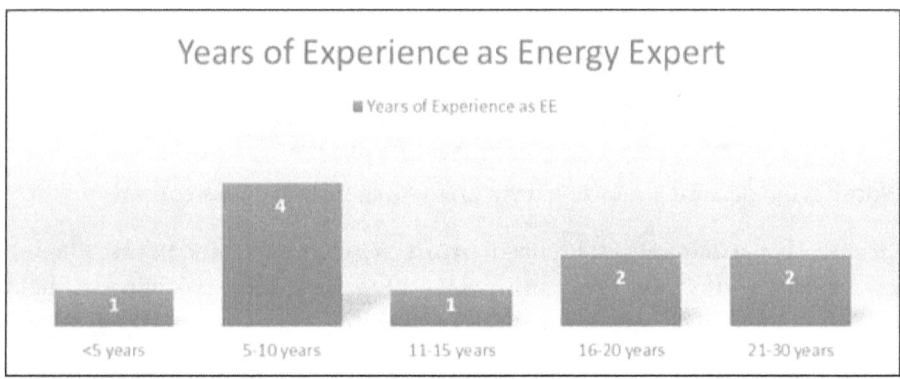

Figure 18 Summary of Experience as Energy Expert. Source: Figure derived by author from analysis of primary data collection.

DETAILS OF ANALYSIS AND RESULTS

After carefully collecting research data, the next necessary step was to perform data analysis. As stated by Creswell (2014), data analysis involves extracting a clear understanding from the information collected from the experiences of participants to achieve an in-depth understanding of the subject matter for the study under consideration (Creswell, 2014). The researcher maintained focus throughout the data collection process by continuously reflecting on the information extracted from the interviewees, by making inferences, and by taking analytical notes to seek answers to answer the research question in this study (Saldana & Leavy, 2011). The analysis of interview data involved preparing and organizing the raw information. To start with, the researcher read through all interview transcripts to gain a general sense of the information collected and to achieve an overall understanding of the perceptions and experiences from the perspective of the participants as it relates to the research question of this study (Creswell, 2014). Following this, a detailed analysis was carried out by deploying a coding process whereby interview data was organized into segments, and the emerging categories were labeled using terms. For the creation of word clouds and the frequency of words, NVivo 11 software was used; however, a manual analysis of the text was performed to derive the theme and sub-themes based on the research questions. These are tabulated in Table 7.

In this chapter, verbatim quotes from the transcripts were used, and then after analysis, the results were classified into various themes, which were clearly presented in the research report. This method was the most suitable way because the topic of the research required data to be collected through interviews of respondents who are directly involved within CIEC of the Klang Valley and was selected following convenience sampling. Presentation of data in this research report includes examples describing the perception of the participants and the attitudes of CIEC of the Klang Valley toward the adoption of energy-efficient (EE) technologies.

The quotations from the transcripts were used to illustrate the themes that emerged from the transcripts. In the presentation of themes, an overview of topics was described followed by verbatim quotes from the transcripts, which were used to illustrate themes that emerged. All quotes extracted from the participants for inclusion in this chapter are verbatim from the transcripts. Where quotations from different participants are incorporated, these were separated by a space, and each quote from each individual research participant was preceded with a unique code allocated to them, such as

P1:P2: etc., that was assigned to them at the start of their statement. Any information that identified the participants was deleted or changed so as to ensure the confidentiality of the interviewee was maintained. Any grammatical errors encountered in the transcripts from the participants' speech were not corrected; the transcripts were presented as spoken by the participants.

Table 7 Research Study Themes and Sub-Themes

Research Questions	Themes	Sub-Themes
Policy and Regulatory:		
RQ#1. Why are electricity consumers in Klang Valley resistant to adopting energy-efficient postures?	**Theme #1** There is insufficient support in terms of meeting policy and regulations requirement, and the policies and regulations are ineffective	**Sub-theme #1a** Governmental non-support and ineffectiveness of policies and regulations **Sub-theme #1b** Governmental support and effectiveness of policies and regulations
	Theme #2 Implementation of environmental policies in regard to reduction of GHG or CO_2 has been neglected	**Sub-theme #2a** Those who do not implement environmental policies **Sub-theme #2b** Those who implement environmental policies
	Economic, Finance, and Marketing:	
	Theme #3 Financial recognition for EE projects and financial risk faced in EE implementation	**Sub-theme #3a** High-risk EE projects which do not get financial recognition **Sub-theme #3b** Low-risk EE projects which get financial recognition

Research Questions	Themes	Sub-Themes
	Theme #4 Availability of finances for EE training and the consumers who are able to afford in-house energy personnel **Behavioral, Information, and Technical:**	**Sub-theme #4a** Non-availability of finances for EE training **Sub-theme #4b** Availability of finances for EE training **Sub-theme #4c** CIECs who can afford in-house EE training
	Theme #5 Awareness of Energy Management techniques among top managers and employees needs to be increased	**Sub-theme #5a** Non-Awareness of EE among top management **Sub-theme #5b** Awareness of EE among top management **Sub-theme #5c** Non-awareness among the employees **Sub-theme #5d** Awareness of employees
RQ#2. How can energy efficiency technologies be implemented in the Klang Valley in Malaysia?	**Theme #6** CIECs who have standard EnMS implemented and integrated into their overall management system	**Sub-theme #6a** CIECs of the Klang Valley who have not implemented EnMS **Sub-theme #6b** CIECs of the Klang Valley who have implemented EnMS **Sub-theme #6c** CIECs of the Klang Valley who have not integrated EnMS in their overall management system **Sub-theme #6d** CIECs of the Klang Valley who have integrated EnMS in their overall management system

Research Questions	Themes	Sub-Themes
	Theme #7 Organizations offering collaboration for EE implementation are limited	**Sub-theme #7a** Governmental and non-governmental organization **Sub-theme #7b** Private organizations and others
	Theme #8 Implementation of CDM and BATs	**Sub-theme #8a** Implementation of CDM **Sub-theme #8b** Implementation of BATs
	Theme #9 There is lack of EE certification on facilities, facility equipment monitoring, and online execution systems	**Sub-theme #9a** Those who have EE certification for their facility **Sub-theme #9b** Those who have facility equipment monitoring system **Sub-theme #9c** Those who have online execution system
RQ#3. How do consumers find energy-efficient technology useful?	**Theme #10** EE implementation provides economic gains and energy cost reduction	**Sub-theme #10a** Economic gains achieved **Sub-theme #10b** Energy cost reduction
	Theme #11 Implementation of EE ensures energy security by mitigation of GHG and CO_2 levels to provide social responsibility (CSR)	**Sub-theme #11a** Energy security and mitigation of GHG and CO_2 **Sub-theme #11b** Local CO_2 and GHG reduction and Social Responsibility

Note: Table derived by author from analysis of primary data collection.

PRESENTATION OF DATA

The main purpose of the research study was to address the primary research questions.

RQ#1: Why are electricity consumers in Klang Valley resistant to adopting energy efficient postures?

RQ#2: How can energy efficiency technologies be implemented in Klang Valley, Malaysia?

RQ#3: How do consumers find energy-efficient technology useful?

While carrying out the interviews for this study, the perspectives and experiences captured from each individual participant on the issue relating to the primary research questions under study were found to be unique. The researcher was also able to gain further knowledge relating to the research question that the researcher did not encounter while carrying out literature research. These experiences are captured and documented later in this section. However, common themes relating to the adaptation of energy-efficient technologies phenomenon emerged within the group. The study revealed the following theme and sub-themes based on the research questions:

RQ#1: Resistance to Adopting Energy-Efficient Postures?

The Themes #1, #2, #3, #4, and #5 and their respective sub-themes are interrelated to address RQ#1.

Theme #1: Policy and Regulatory. There is insufficient support in terms of meeting policy and regulations requirements, and the policies and regulations implemented are ineffective.

The research participants' responses on the different support attained to meet the required policies and regulations in EE which are provided to the CIECs of Klang Valley through local agencies or firms responsible for EE and the effectiveness of the policies and regulation are summarized below in Table 8:

Table 8 Participants' Responses: Theme #1

Participant's Code	Response to: Support in Terms of Meeting Policy and Regulatory and Effectiveness of the Policies and Regulation
P1	Policy and regulation are in place by Energy Commission of Malaysia but not much enforcement.
P4	Support in EE investments and tax exemption.
P6	More on awareness and not on enforcement and no achievement target provided.
P10	Local agencies, organizations don't directly provide the support.

Note: Table derived by author from analysis of primary data collection.

According to the support in terms of meeting policy and regulation, the Energy Commission of Malaysia has imposed an act titled the Malaysian Regulation of Electrical Supply Act 1990 under the section of Efficient Management of Electrical Energy Regulation (Energy Commission of Malaysia, 2008). This act states that a person responsible for the operations of a building or facility which receives or generates a total net electrical energy equal to or exceeding 3,000,000 kWh over any period not exceeding six consecutive months will be imprisoned for a period of no less than one year, or fined Ringgits Malaysia Five thousand, or both for if he or she does not abide by the act. Since no person has been fined or imprisoned up until now, it is evident that there is no strict enforcement in Malaysia of the Act and financial support is limited (P1, P2, P3, P5, P7, P8, P9). The majority of Malaysian know about this, especially the energy managers in Malaysia, who do not get much support from the Energy Commission. The Energy Commission is good with their policy-making and enforcing other regulatory requirements. There is some training provided by local private entities engaging energy experts from Singapore, and this might not be sufficient or may not be relevant to Malaysian needs. The Energy Commission sets the regulations; however, it takes a concerted effort from all levels, including consumers, associations, NGOs, and the rest of the people, to work together. This is the most important aspect to ensure that the policies and regulations are effective. Usually, these policymakers provide a lot of information, guidelines, and recommendations over the internet through their web pages. Yet all of this is rarely taken up by the consumers.

This information is always available (on websites and published documents) if one actually seeks for it. It is there but usually the effort to seek this information by the CIECs is absent (P8).

Sub-Theme #1a – Governmental Non-Support and Ineffectiveness of Policies and Regulations. The non-support in terms of meeting policies and regulations implemented by the Energy Commission of Malaysia and the non-effectiveness of the policies and regulation as claimed by the research participants are as follows.

> **P1:** "Policy and regulation are in place by Energy Commission of Malaysia, but the thing is there is not much of enforcement by the commission, so CIEC is not keen to be part due to lack of enforcement."
>
> **P2:** "Main policy and regulations are made by the Energy Commission of Malaysia, but there is no strict enforcement by the commission."
>
> **P5:** "Here the main policy comes from the Energy Commission of Malaysia, but there are no strict regulations in Malaysia."

Sub-Themes #1b – Governmental Support and Effectiveness of Policies and Regulations. The following research participants claim there is support in implementing policies, and the policies and regulation implemented by the Energy Commission of Malaysia are effective on the CIECs of Klang Valley.

> **P4:** "There was a support in terms of financial scheme on EE investments as well as tax exemptions… No clear guidance from the government or local authority as far as EE is concerned. So as far as CIEC is concerned, meeting policy or regulatory requirement is not a major item in their business requirement or even commercial objectives."
>
> **P6:** "Regulations on EE is by Energy Commission but more on awareness and not on enforcement and no achievement target provided on EE for any reduction."

Theme #2: Implementation of Environmental Policies in Regard to Reduction of GHG or CO_2 has been Neglected. The research participants' responses on environmental policies that are being implemented by the CIECs of Klang Valley to reduce GHG or CO_2 are summarized in Table 9:

Table 9 Participants' responses: Theme #2

Participant's Code	Response to: Environmental Policies in Regard to Reduction of GHG and CO_2
P1	Larger Multinational Companies (MNCs) incorporate CSR (corporate social responsibilities) policies and mostly absent in smaller and local companies.
P4	Mostly utilize what is stipulated in current ISO standards.
P5	Larger MNCs incorporate CSR policies and mostly absent in smaller and local companies.
P6	Utilize ASEAN Energy Management Scheme (AEMAS) or International Standards Organization (ISO) 50001 standards.

Note: Table derived by author from analysis of primary data collection.

Many large multinational companies have incorporated corporate social responsibility (CSR) policies for the reduction of GHG and CO_2, but small local companies do not have strong policies on the GHG and CO_2 reduction among the CIEC of Klang Valley (P1, P2, P3, P5, P7). The more kWh you save, the less CO_2 emission, which means one kWh saved will reduce CO_2 emission by 0.0007 tons (P10). There are few CIEC clients, mainly MNCs, who have implemented energy management/savings initiatives and policies to reduce CO_2 and GHG, except for the local or smaller companies since there is no specific policies or regulations requiring them to implement these initiatives and policies.

Sub-Theme #2a – Those who do not Implement Environmental Policies. The following research participants claim that the following CIECs of Klang Valley do not implement environmental policies.

P1: "Smaller and local companies tend not to have such policies."

P3: "Smaller and local companies do not have strong policies."

P5: "Local companies I have not seen anybody implement this."

Sub-Theme #2b – Those who Implement Environmental Policies. The following research participants assert that the following CIECs of Klang Valley implement environmental policies.

P1: "Larger MNCs have incorporated such as CSR policies in hope to contribute to the reduction GHG and CO_2."

P9: "Many listed companies with capitalization are required to reduce their carbon footprint or CO_2, and they have CO_2 reduction roadmap with yearly action items and monitoring carried out and reported."

P10: "My CIEC clients have implemented energy management/savings initiatives and policies to reduce CO_2 and GHG."

Theme #3: Economic, Finance, and Marketing: Financial Recognition for EE Projects and Financial Risk Faced in EE Implementation. The research participants' responses on EE projects that get financial recognition and the different financial risks faced after EE projects have been implemented are summarized in Table 10:

Table 10 Participants' Responses: Theme #3

Participant's Code	Response to: Financial Recognition for Types of EE Projects	Response to: Financial Risk
P1	Rapid payback period or high ROI or low investment cost	Non-delivery within the stipulated period and expected ROI not being met.
P4	High returns in shortest time	Longer ROIs or system malfunction affecting targeted savings value is not achieved.
P7	Rapid payback period and high return on investment (ROI)	EE might affect corporation core business and also a possibility of failing in delivering expected ROI or payback.
P9	Any Energy Performance Contracting (EPC)	No financial risk if implemented after a thorough energy audit.

Note: Table derived by author from analysis of primary data collection.

The type of EE projects that gets financial recognition among CIEC of Klang Valley are rapid payback period, high ROI, low investment costs, high returns in a short period of time, and any energy performance contracting (EPC) (P1, P2, P3, P4, P6, P7, P8). In fact, the implementation of energy efficiency projects might affect the operation of CIEC core business processes, which is one of the main concerns that may compromise productivity. Hence, the types of financial risks that are faced by the clients of CIEC of Klang Valley while EE projects are being implemented are the non-deliverance of the expected savings within the stipulated time period; thus, the return on investment is not being met, which leads to financial risk. When tariffs are being revised, the calculations on the payback and ROI will differ, leading to financial risks (P1, P2, P3, P4, P5, P6, P7, P8, P10). Financial risk is not foreseen if the project is implemented after a thorough energy audit has been carried out (P9).

Sub-Theme #3a – High-Risk EE Projects which do not Get Financial Recognition. The following research participants claim that the following types of high-risk EE projects do not gain financial recognition.

- **P1:** "Main risk in EE projects are non-deliver of the expected saving in the stipulated period (within an agreed payback period) and the expected ROI not being met."
- **P2:** "Implementation might affect the operation of their core business processes; the project might fail in delivering the expected ROI or payback period."
- **P3:** "Implementation of EE project might affect the operation of their core business Processes, and the EE project might fail in delivering the expected ROI or the payback period."
- **P4:** "Longer ROIs or system malfunction that caused targeted savings value is not achieved."

Sub-Theme #3b – Low-Risk EE Projects with Financial Recognition. The following research participants claim that the following types of low-risk EE projects gain financial recognition.

- **P1:** *"The EE project will get financial recognition are the ones having rapid payback period or high ROI or low investment cost."*

P6: *"Lower payback period (less than 2 years) with life span more than 5 to 10 years."*

P9: *"Any EPC offering gets financial recognition as there no upfront cost to the CIEC."*

Theme #4: Availability of Finances for EE Training and Clients who can Afford in-House Energy Personnel. The research participants' responses on the various types of finances accessible for EE training toward in-house personnel and the CIECs of Klang Valley who can afford in-house energy personnel are summarized in Table 11:

Table 11 Participants' Responses: Theme #4

Participant's Code	Response to: Type of Financials Available	Response to: Affordable Clients
P1	Human Resource Development Funds (HRDF)	MNCs
P4	Green Technology Financing Scheme (GTFS)	Large-scale companies, diversified companies
P6	There is not specific for EE but can 100% refund through HRDF	MNCs and large-scale companies with HRDF registered
P8	There are some fundings available under MIDA	Companies that consume more than 3 million kWh

Note: Table derived by author from analysis of primary data collection.

There are insufficient financial funds of any kind for the CIEC of Klang Valley to pay for EE training for their in-house personnel (P2, P3, P7, P8), but in the past, there were some funds available under the Malaysian Industry Development Agency (MIDA) worth about RM200,000 to train the members of the Malaysian Energy Professional Association (MEPA) and many others from the institution (P8); Unfortunately, the funding has since ceased. Other funding is jointly available from UNIDO, a five-year program that just ended in 2016, which provided free training for people from the industrial sector who showed evidence of commitment from the authorities (P8). Some of the external funding available were from Energy Services Companies (ESCOs) that provide some financing for EE training

(P10). There were funds available from the Malaysian government agencies for general purposes only and that was limited. Some funding was also made available by non-governmental organizations (NGO) and from the Malaysian government's Human Resource Development Fund (HRDF). However, the 100% refund through HRDF that is still available is the Green Technology Financing Scheme (GTFS) (P4). The types of clients that could afford in-house energy personnel are mostly large multinational companies (MNCs) or big energy consumers and not necessarily MNCs; national companies which consume more than three million kWh (kilo-Watts-hour) and above over a period of six months are the ones who will definitely see the benefit of having in-house energy personnel (P1, P2, P3, P4, P5, P6, P7, P8, P10). Consumers who have exceeded the 500 thousand kWh in any six months period engage the services of Registered Electrical Energy Managers (REEM) energy personnel due to regulatory requirements (P8, P10).

Sub-Theme #4a – Non-Availability of Finances for EE Training. The following research participants assert that there is limited or no funding available for in-house training for CIECs of Klang Valley.

> **P1:** "Unfortunately, there are not much financing available for EE training except for the usual HRDF, which are not specifically aimed at EE."
>
> **P2:** "There are not much financing available for EE training… Fund available from the government for general purpose training, which is limited, and for the in-house personnel is usually decided by the management."
>
> **P3:** "There are not much financing available for EE training… Fund available are from government agency such as HRDF for general purpose training, which are limited."
>
> **P7:** "There isn't much financing available… Sometimes they got the funding from NGO organization or from the government HRDF."

Sub-Theme #4b – Availability of Finances for EE Training. The following research participants state that there is funding available for in-house training for CIECs of Klang Valley.

> **P6:** "There is but not specific for EE but can have 100% refunds through HRDF."
>
> **P5:** "It is only through contribution."

P8: "Actually, there was some fundings to be tapped under the Malaysian Industry Development Agency (MIDA) which I am aware."

P10: Only external funds such as ESCOs or Energy Services Companies that provides some financing for EE training."

Sub-Theme #4c – CIECs who can Afford in-House EE Training. The following research participants claim that the CIECs of Klang Valley who can afford in-house training are as follows.

P2: "Large multinational companies as well companies that utilize high energy."

P3: "Again, only large MNCs can afford in-house energy personnel."

P4: "Large-scale companies, diversified companies.... A huge company with multiple industries, properties, construction, and facility management."

P10: "Clients who have exceeded the 500 thousand kWh in any six months period engaged the services of in-house energy personnel (REEM) due to regulatory requirements."

Theme #5: Behavioral, Information, and Technical: Awareness of energy Management Techniques Among Top Managers and Employees Needs to be Increased. The research participants' responses on the level of awareness regarding EE among top managers and employees according to the type of CIEC of Klang Valley are summarized in Table 12:

Table 12 Participants' Responses: Theme #5

Participant's Code	Response to: Top Manager Awareness	Response to: Employees Awareness
P1	Top managers are fully aware in MNCs; local and smaller firms lack in awareness	Employees in large firms are aware; local and small firm employees lack in awareness.
P4	Awareness is fairly low	Awareness among employees are higher than top management.

Participant's Code	Response to: Top Manager Awareness	Response to: Employees Awareness
P5	Fully aware	Employees in large firms are aware; smaller firm employees lack in awareness.
P6	Low awareness	Firms with energy management have proper way to cover awareness to all employees.

Note: Table derived by author from analysis of primary data collection.

In large multinational companies, the members of top management are fully aware of energy efficiency (EE) and priority is given to EE as part of their corporate social responsibility (CSR) (P1, P2, P3, P5, P7, P8, P9). Whereas in local, smaller, and medium enterprises (SMEs), there is a lack of concern for EE even though top management is aware of EE, but due to tight profit margins and high competition among SMEs of a particular industry, the top management of SMEs tend to focus on gaining profit by focusing on the business revenue rather than implementing EE. Furthermore, SMEs are unable to afford the capital cost of implementing EE measures (P4, P6, P10).

Awareness among employees regarding EE implementation is higher among the larger multinational companies since they conduct regular, internal energy awareness programs within their organizations to make employees become aware of EE (P1, P2, P3, P4, P5, P7, P9). Some innovative methods used to create awareness of EE in an organization are to encourage the involvement of employees in energy consciousness, energy awareness campaigns, and by holding competitions between different departments within an organization to determine the most effective energy-efficient solutions and solutions that provide the most innovative ideas for energy conservation. These kinds of campaigns and competitions will cost money, and most of the smaller and local companies cannot afford these kinds of campaigns in order to promote employee awareness (P2, P3, P7, P9).

Sub-Theme #5a – Non-awareness of EE among top management. The types of CIECs of Klang Valley whose top management do not have awareness in EE as stated by the research participants are as follows:

P1: "Lack of awareness among the local and smaller organizations' topmanagement."

P2: "Awareness in EE is lacking in local and smaller organizations."

P8: "Small and medium companies are firefighting to make the profit."

P10: Almost none... They don't want to know about it."

Sub-Theme #5b – Awareness of EE among Top Management. The types of CIECs of Klang Valley whose top management have awareness in EE as asserted by the research participants are as follows:

P1: "MNC's top management are fully aware."

P2: "The management is fully aware in MNCs."

P3: "Top management in large-scale organization such as MNCs are fully aware."

P7: "In large MNC the top management are fully aware."

P8: "Large companies, top management are fully aware."

Sub-Theme #5c – Non-Awareness among the Employees. The CIECs of Klang Valley whose employees do not have awareness in EE as claimed by the research participants are as follows:

P1: "Local and smaller companies most of the employees are not aware."

P2: "Campaign and competition will cost money, and most of the smaller and local companies cannot afford this kind of campaign to make their employee aware."

P8: "In terms of awareness, it is more on general encouragement only. It is not more of an enforcement compared to people in Japan, where the energy price is so high.

They will not hesitate to reprimand their colleague who is wasting energy. However, in this region, even the slightest note will become offensive."

P9: "Campaigns/competitions are absent in SMEs and employees, and naturally, their awareness will always be low."

Sub-Theme #5d – Awareness of Employees. The CIECs of Klang Valley whose employees have awareness in EE as asserted by the research participants are as follows:

P1: "Employees of large organizations are very aware."

P2: "Larger multinational companies conduct regular energy conservation campaign to make employee awareness... Campaign and competition will cost money."

P6: "Only companies with energy management in place will have a proper way to cover the awareness on EE to all employees."

P10: "Employees could be engineers, mechanical, or electrical engineers. Yes. I'm sure they are aware of it."

RQ#2: How Can Energy Efficiency Technologies be Implemented?

The themes #6, #7, #8, and #9 and their respective sub-themes are interrelated to address RQ#2.

Theme #6: Small Electricity Consuming Companies Cannot Afford to Implement EnMS and Integration into the Overall Management System of the Company. The research participants' responses on the standards of how EnMS was implemented and its integration into their overall management system among the CIEC of Klang Valley are summarized in Table 13.

Table 13 Participants' Responses: Theme #6

Participant's Code	Response to: Standard EnMS Implemented	Response to: Integration of EnMS
P1	MNCs implement ISO 50001. Smaller local firms do not imply	Local or smaller firms do not integrate EnMS due to non-affordability and larger firms integrate.
P3	Large MNCs implemented ISO 50001	Large local and MNCs integrate EnMS; smaller firms do not integrate EnMS.
P4	ISO 50000 series	EnMS are not integrated into other management systems.
P5	MNCs go for ISO 50001	Large and MNCs implement EnMS; smaller firms do not integrate EnMS.

Note: Table derived by author from analysis of primary data collection.

The standard energy management system implemented by the CIEC of Klang Valley is ISO 50001, but not all CIEC of Klang Valley have implemented energy management systems (P1, P2, P3, P4, P5, P6, P7, P8, P9, P10). Mostly large MNCs have implemented energy management systems of a known international standard, and so, most of these MNCs have implemented ISO 50001 standard for energy management (P1, P2, P7, P8). Some of the reasons for the other organizations that have not implemented an energy management system have to do with the high cost associated with it. The integration of the energy management system (EnMS) into an overall management system is implemented mostly in large organizations such as large multinational organizations (MNCs). Even though it is costly, it is incorporated in the overall organization, whereas it is not implemented in the small organizations due to non-affordability, and hence, it is not integrated into smaller organizations (P1, P2, P5, P7).

Sub-Theme #6a – CIECs of the Klang Valley who have not Implemented EnMS. The types of CIECs of Klang Valley that have not implemented EnMS as stated by the research participants are as follows:

P1: "Smaller local companies due to cost they tend not to implement."

P10: "None of my clients have implemented the ISO 50001, but whatever policies and procedures and training and documentation and quality management systems I have done for them, is basically following the said standards 50001."

Sub-Theme #6b – CIECs of Klang Valley who have Implemented EnMS. The types of CIECs of Klang Valley who have implemented EnMS as asserted by the research participants are as follows:

P1: "MNCs implement a lot of standard for energy management system such as ISO 50001."

P2: "Large MNCs have implemented energy management system according with international standard have already implemented ISO 50001."

P5: "There isn't much of a standard that I have seen except for multinational companies which go for ISO 50001."

P6: "ISO 50001, AEMAS, UNIDO EnMS."

Sub-Theme #6c – CIECs of Klang Valley who have not Integrated EnMS in their Overall Management System. The types of CIECs of the Klang Valley who have not integrated EnMS in their overall management system as claimed by the research participants are as follows:

P1: "I would say local or smaller organizations do not integrate EnMS due to non-affordability."

P2: "Smaller organizations do not implement due to affordability, so they are not integrate."

P4: "Mostly EnMS systems are not integrated into the other management system."

P7: "Smaller organizations cannot afford to implement EnMS and obviously do not integrate EnMS into their overall management system."

Sub-Theme #6d – CIECs of Klang Valley who have Integrated EnMS in their Overall Management System. The type of CIECs of Klang Valley who have integrated EnMS in their overall management system as claimed by the research participants are as follows:

P1: "Larger organizations integrate EnMS into the overall management system."

P2: "Mostly carried out by large or multinational companies which is costly but are incorporated in the organizations overall."

P7: "EnMS implementation mostly implemented by large MNCs integrate EnMS into their overall management system."

P10: "Okay, as I said earlier, the energy management system which I follow is an extract of ISO 50001. So, any documentation, any training, any policies, any QMS (Quality Management System) which I implement for my CIECs is from them."

Theme #7: Organization Offering Collaboration for EE Implementation are Limited. The research participants' responses to the types of organizations offering collaboration for EE implementation with CIEC of Klang Valley are summarized in Table 14.

Table 14 Participants' Responses: Theme #7

Participant's Code	Response to: Collaboration with Other Organization in EE Implementation
P2	Absence of government or non-government support. Privately owned energy service company offers low-lying energy solution.
P4	Collaboration with other EE industrial players through meeting, conferences, energy performance update, and info-sharing.
P9	KeTTHA and MGTC have collaborated with some through joint funding initiatives.
P10	Private company that is ESCO.

Note: Table derived by author from analysis of primary data collection.

CIEC of Klang Valley are not receiving collaboration nor support from the government and non-government organizations (NGO) for implementing EE action plans and EE programs. But many private organizations are helping in implementing EE action plans and EE programs (P1, P2, P3, P7, P8, P10). Some of the few mentioned private organizations helping in the EE implementation are energy service companies (ESCOs) and the Malaysian Energy Professional Association (MEPA). Both offer low-cost energy services and solutions, helping in EE action plans and services (P1, P2, P3, P5, P7, P8, P10).

Sub-Theme #7a – Governmental and Non-Governmental Organizations. The types of governmental and non-governmental organizations that provide collaboration with CIECs of Klang Valley for the implementation of EE as claimed by the research participants are as follows:

P1: "No government or non-government organizations assisting."

P2: "Not been any government or non-governmental organizations (NGOs) organizations assisted."

P6: "Only available for commercial buildings are to apply for energy audit grant through SEDA for industrial facilities they can apply to MGTC... No guarantee the grants will be approved due to limited available funds."

P8: "Not been any government or known organizations assisted… KeTTHA and MGTC have sporadically collaborated with some CIEC with some joint funding initiatives, but this has yet to scale-up."

Sub-Theme #7b – Private Organizations and Others. The types of private organizations providing collaboration with CIECs of Klang Valley for the implementation of EE as stated by the research participants are as follows:

P1: "Privately owned Energy Service Company (ESCOs) that offer low-lying energy solution."

P2: "Energy Services Companies (ESCOs) which are privately owned, offering energy solutions which are easily implemented."

P4: "Collaboration comes in the forms of meeting, energy performance update, and info-sharing as well as conferences with other EE Industrial player."

P8: "Private owned but nevertheless the MEPA as core companies… Doing this on a regular basis have been offering low-lying energy solutions."

Theme #8: Failure of CDM has Led to a Lack of Monitoring in CO_2 Reduction. The research participants' responses on Clean Development Mechanism and the types of best available technologies implemented by the CIEC of the Klang Valley are summarized in Table 15.

Table 15 Participants' Responses: Theme #8

Participant's Code	Implementation of CDM	Types of BATs Implemented
P4	Not sure about the level of implementation of CDM in Klang Valley	LED lighting system.
P5	No implementation of CDM	LED lighting system and VFDs for motors.
P6	No longer available	LED lighting system.
P7	No funds left in Carbon Trust and CDM process had fizzled out	LED lighting system, intelligent chiller control, and power optimizer.

Note: Table derived by author from analysis of primary data collection.

Under the Clean Development Mechanism (CDM) (UNFCCC, 2006) guidelines, emission-reduction projects in developing countries can earn certified emission reduction credits. These saleable credits can be used by industrialized countries to meet a part of their emission reduction targets as outlined in the Kyoto Protocol (UNFCCC, 1998). Governments then issue permits up to the agreed limit, and these are either given free or auctioned to companies in the sector. They work by setting an overall limit or cap on the amount of emissions that are allowed from significant sources of carbon, including the power industry as well as automotive and air travel, at regional, national, and international levels. If a company curbs its own carbon significantly, it can trade the excess permits on the carbon market for cash. If it is not able to limit its emissions, it may have to buy extra permits. Therefore, the carbon credits could be traded and sold and utilized by industrialized nations to purchase the carbon credit required to pay for their excess CO_2 emission under the Kyoto Protocol requirement. Creating a market in something with no intrinsic value such as carbon dioxide is very difficult. There is a need to promote scarcity, which should strictly limit the right to emit CO_2 so that it can be traded. Due to this, a glut of permits was created. These permits have often been given away for free, which has led to a collapse in the price, and no effective reductions in emissions compounded, allowing offset permits gained from paying for pollution reductions in poorer countries to be allowed to be traded as well (Kill, Ozinga, Pavett & Wainwright, 2010). Due to the above-mentioned issues in carbon trading, and from what observation shows, the CDM process has died off in Klang Valley (P1, P2, P3, P5, P6, P7, P8, P9, P10).

The best available technologies (BATs) that can commonly be found among the CIEC of the Klang Valley clients are LED lighting system, low energy consuming motors, VFDs for motors, intelligent chiller control, power optimize, inventor-type air-conditioning, air-water system, and high-efficiency motors (P4, P10). But most of all, the best available technology is the LED, which is very straight forward; the implementing time is very short and is completely manageable cost-wise (P1, P2, P3, P4, P5, P6, P7, P8, P10).

Sub-Theme #8a – Implementation of CDM. The implementation of CDM by the CIECs of Klang Valley as claimed by the research participants are as follows:

> **P1:** "Worth noting that there are not much funds left in the Carbon Trust, and my observation shows that the CDM process has died."

P2: "Carbon trade has fizzled out as there are insufficient funds remaining in the Carbon Trust and a lots of developing countries have stopped exercise this trading."

P4: "CIEC in Klang valley were not adequately exposed toward clean environment... Not very regular at this point, but if you look at specific regulation like protocol, basically they are in topic on air-conditioning system."

P5: "I don't think any of these have been implemented here... There was talk that they wanted to implement CDM through CO_2 emission control and carbon trade, but specifically, there is nothing much that has been done on this."

Sub-Theme #8b – Implementation of BATs. The types of BAT implemented by the CIECs of Klang Valley as stated by the research participants are as follows:

P1: "The BAT energy saving lighting system."

P2: "BATs commonly can be found are LED lighting system and high-efficiency motors."

P4: "BATs is usually with the lowest cost, with the shortest period of returns, and the simplest to implement... LED it is very straightforward, implementation time is very short, and completely manageable cost-wise."

P5: "BATs are LED lighting, VFDs for motors... Building automation system generally, you are able to manage the facility system according to the demand."

Theme #9: There is a Lack of EE Certification on Facilities, Facility Equipment Monitoring System, and Online Execution System. The research participants' responses on the various types of EE certification on facilities, types of facilities equipment monitoring system employed, and the types of online execution system adopted by the CIEC of Klang Valley are summarized in Table 16.

Table 16 Participants' Responses: Theme #9

Participant's Code	Response to: Those Who Have EE Certification on Their Facility	Response to: Those Who Have Facility Equipment Monitoring System	Response to: Those Who Have Online Execution System
P1	Leadership in Energy and Environment and Design (LEED) and Green Building Index (GBI)	Online real-time computer-based on monitoring system	Building Automation System (BAS) and Building Energy Management System (BEMS)
P2	LEED and GBI	Large-scale: Online real-time computer-based on monitoring system Small-Scale: Human observation and manual data recording	Large organizations: BAS, Building Intelligent System (BIS), and BEMS Small organizations: Manual logging via employee experience
P4	Local certification: PH green certification, GreenPASS, and MyCrest. International Recognition: LEED and Greenmarks	Building Automation System (BAS)	BAS

Participant's Code	Response to: Those Who Have EE Certification on Their Facility	Response to: Those Who Have Facility Equipment Monitoring System	Response to: Those Who Have Online Execution System
P5	LEED, GBI, and CDM	Large-scale: Real-time systems such as Building Information Management System (BIMS). Small-scale: No integrated system for only specific equipment	BAS and BEMS

Note: Table derived by author from the analysis of primary data collection.

Among the CIEC of Klang Valley facilities that have been recognized as energy efficient, the majority have been certified through the LEED (Leadership in Energy and Environmental Design) and GBI (Green Building Index) certification standards (P1, P2, P3, P4, P5, P6, P7, P8, P9). The local certification is through the Penarafan Hijau (PH – Green Certification), GreenPASS, and Malaysian Carbon Reduction Assessment Tool (MyCrest) (P4), whereas international recognition comes from the American Leadership in Energy and Environment Design (LEED) and Greenmarks (Singapore).

Facility equipment monitoring at CIECs of Klang Valley is generally carried out through the online real-time computerized monitoring system, which is through field gauges, meters, sensors. Other data acquisition devices are carried out by utilizing Building Automation System (BAS), Building Intelligent System (BIS), and Building Energy Management System (BEMS) (P4, P5, P6). These systems are generally being utilized by large-scale industries, whereas human observations, manual recordings, and checklists are generally utilized by small-scale industries (P2, P3, P5, P7, P9).

The type of online execution system available to the end-user via user interfaces, which helps manage, monitor, and control energy usages being utilized by the CIEC Klang Valley, are Building Automation System (BAS), Building Intelligent System (BIS), and Building Energy Management System (BEMS) that integrated with customized software for specific engineering system. These systems are usually utilized by larger organizations because they are expensive. The smaller organizations utilize manual logging and assessment on a regular basis due to affordability (P1, P2, P3, P4, P5, P6, P7, P8, P9, P10).

Sub-Theme #9a – Those who have EE Certification their Facility. The types of EE certification on CIECs of Klang Valley clients' facilities as claimed by the research participants are as follows:

P1: Only when such facilities complied let says with LEED or GBI, but majority are not been certified to known certification standard."

P2: "Some of the larger facilities are certified in LEED Energy Efficiency and Environmental Design and also some of the new facilities are certified by GBI; majority are not been certified to any know certification standard."

P4: "There are local as well as international energy efficiency certifying bodies… Local Certification could be from Penarafan Hijau (PH – Green Certification),GreenPASS, and Malaysian Carbon Reduction Assessment Tool (MyCrest)… International Recognition could come from American LEED and Greenmarks (Singapore)."

P5: "I don't see anything specific to EE, but to some larger extent, LEED, GBI, and CDM building system… Housing system in Malaysia has another system. In total, they are about six green rating systems in Malaysia you can use to establish your EE requirement."

Sub-Theme #9b – Those who have Facility Equipment Monitoring System. The types of facilities equipment monitoring system employed by CIECs of Klang Valley clients' facilities as stated by the research participants are as follows:

P1: "The clients monitored their facilities via online real-time computer-based on monitoring system which has facilities such as field gauges, meters, and sensor. In small-scale organization, mostly through human observation."

P2: "Clients monitored via online real-time computer-based on monitoring system… such as field gauges, meters, and sensors. Large organizations monitor via online real-time computer-based monitoring through field gauges, meters, and sensors, and in small-scale organizations usual through human observation and manual data recording."

P3: "Large organizations monitor via online real-time computer-based monitoring through field gauges, meters, and sensors, and in small-scale organizations, mostly through human observation and checklists."

P5: "Usually through real-time system users of building management information system through gauges, meters, sensors… Smaller scale probably it is not an integrated system but specific like chilling will require to manage the energy demand."

Sub-Theme #9c – Those who have Online Execution System. The types of online execution system adopted by CIECs of Klang Valley clients as asserted by the research participants are as follows:

P1: "Such online usages system mainly BAS, BEMS they are used to monitor and control energy usages."

P2: "Large organizations utilize system such as BAS, BIS, and BEMS to monitor and control energy usages… For smaller organizations, the energy is monitored via manual logging and control via employee experience."

P4: "It will be BAS but connected via online system and online technology."

P7: "Large organization utilizes BIS or probably BEMS to control their energy usage… Smaller organization the energy is monitored via manual logging and controlled through the human actions."

RQ#3: How Do Consumers find Energy-Efficient Technology Useful? The themes #10 and #11 and their respective sub-themes are interrelated to address RQ#3.

Theme #10: EE Implementation Provides Economic Gains and Energy Cost Reduction. The research participants' responses on the economic gains achieved and energy cost reduction brought about by the implementing of EE are summarized in Table 17.

Table 17 Participants' Responses: Theme #10

Participant's Code	Response to: Economic Gains Achieved	Response to: Energy Cost Reduction
P1	Lower cost production and increase overall revenue	Could reduce consumption back to the level of energy consumption before the expansion.
P4	Lower cost production and increase overall revenue	Lower energy cost to the same product yield.
P9	Reduce commercial building energy consumption by 40% compared to a baseline kWh	Energy-efficient measures on the 'big ticket' items consistently will avoid CIECs from going into 'energy shocks.'

Note: Table derived by author from analysis of primary data collection.

It is obvious that the lower cost of production will provide economic gains, which will increase CIEC revenue and profitability, which will lead them to become competitive against others in the similar industry through increased competitiveness as their products can be offered at a lower price (P1, P2, P3, P4, P5, P6, P8). A typical commercial building can reduce its energy consumption by 40% when compared to a baseline kWh (P9). The gains, even though small, cannot be explained in terms of dollars and cents; it is expressed in kWh. With business expansion, it is obvious that the absolute energy consumption will increase, but at the same time, the increase in efficiency of production due to EE will also correspondingly increase, thus increase in revenue; this is effective when the production cost decreases and the revenue increases (P10).

The cost of energy cost is usually on the average of about 20% of the production cost. So, any energy efficiency implementation will only translate to 10% to 20% energy saving, which translates to 2% to 4% reduction in the cost of production (P8). Hence, implementing EE technologies benefits the CIEC of the Klang Valley as it reduces energy cost.

Sub-Theme #10a – Economic Gains Achieved. The economic gain achieved by CIECs of Klang Valley clients through the implementation of EE measures as claimed by the research participants are as follows:

    P1: "Help drive down cost of production and the increase in overall revenue."

P2: "Lower cost of production and the increase in overall revenue."

P9: "A typical commercial building can reduce its energy consumption by 40% (compared to a baseline kWh)."

P10: "There will be gains, but it might be small. That is why sometimes you cannot talk in terms of dollars and cents; you might have to talk in terms of kWh."

Sub-Theme #10b – Energy Cost Reduction. The energy cost reduction by CIECs of Klang Valley clients through the implementation of EE measures as asserted by the research participants are as follows:

P4: "Having EE technologies lower energy cost to the same product yield... Savings realized from the EE measures will then be an offset to the additional amount of energy cost estimated for the planned expansion."

P8: "So, as long as the production cost and the revenue are correlated positively, the energy efficiency will still not be an attractive thing for them to focus on... Because usually the energy cost on average it's about 20% of the production cost. So, any energy efficiency implementation will only translate to 10% to 20% improvements, which translates to 2% to 4% of the production cost."

P9: "A detailed energy audit and coming up with a 5 years roadmap... Better understanding of where most of the expenditure for energy is going to and by implementing energy-efficient measures on the 'big ticket' items consistently will avoid these CIECs from going into 'energy shocks.'"

P10: "Yes, it will help because here your benchmark already your specific energy consumption, and it would also help in the business expansion or increase in production can be pushed into the non-peak hours, so they can enjoy lower energy tariff to overcome the added cost for the business expansion."

Theme #11: Implementation of EE Technologies Ensures Energy Security by Mitigation of GHG and CO_2 Levels to Provide Corporate Social Responsibility (CSR). The research participants' responses on the various energy security measures to improve business productivity and corporate social responsibility (CSR) to reduce the local GHG and CO_2 emission by CIECs of Klang Valley are summarized in Table 18:

Table 18 Participants' Responses: Theme #11

Participant's Code	Response to: Energy Security	Response to: Local CO_2 and GHG Reduction
P1	Maintain their cost of product even when price of electricity increases	Implementing EE indirectly reduces CO_2 and GHG emission while power plant demand is reduced, then natural resources like coal and fossil fuel required to generate power will indirectly reduce CO_2 and GHG emission
P4	Translates to reduced operating costs and reducing the amount of energy used and reducing the emission of CO_2.	Implementing EE indirectly reduces CO_2 and GHG emission. Direct impact in reduced amount on depletion of our natural resources and replanting of trees could be promoted.
P8	Not concerned about energy security but concerned about business productivity, least bothered about the GHG or CO_2.	Applicable to large organizations, and for small and medium organizations, it's not significant. Reduction in power plant will lead to a reduced use of natural resources such as coal and fossil fuel which is used to generate power, and this will indirectly reduce emission of CO_2 and GHG.
P9	Carry out energy audit, coming up with 5-year energy reduction roadmap with the 1st year focusing on the operational low-hanging fruit.	EE measures correctly implemented, Scope 2 of ISO 14064 GHG reduction related to 'purchased electricity' and will reduce and will lead to reduced use of natural resources such as coal and fossil fuel to generate power and reduce emission of CO_2 and GHG.

Note: Table derived by author from analysis of primary data collection.

The CIEC of Klang Valley clients, in order to be cost-effective, have achieved energy security in order to improve business productivity and mitigate GHG and CO_2 reduction by implementing EE measure, enabling them to lower or maintain their cost of product even when the price of energy has increased. The reduction in energy due to EE measures will compensate for the increment in energy price; this further achieves energy security by improving or maintaining business productivity, in turn mitigating GHG and CO_2 (P1, P2, P3, P4, P4, P5, P6, P7). As a further measure, the CIEC of Klang Valley clients should carry out an energy audit and come up with a five-year energy reduction roadmap within the first year by focusing on the operational low-hanging fruit and utilize a part of their savings to reinvest into the subsequent year as EE investments. For example, part of the previous EE saving will pay for the future EE project investment (P9). And finally, it can be cost-effective to achieve energy security by being conscious of the latest development in EE and establish benchmarks in specific energy consumption and compare it against similar industry benchmarks in order to gauge the effectiveness of their energy improvement programs (P10).

CIEC of Klang Valley clients can contribute to further reduction of local CO_2 and GHG emission and can be socially responsible (CSR) by reducing energy consumption through implementation of EE measures; this indirectly reduces the CO_2 and GHG emission and the demand for power generation. When the demand for power generation reduces, natural resources such as coal and fossil fuel to generate power will eventually reduce as each of kWh reduction in electricity will help to reduce CO_2 emission by 0.694 kg of CO_2 (P6). This will directly reduce CO_2 and GHG emission and support corporate social responsibility (P1, P2, P3, P4, P5, P6, P7, P8, P9, P10). The Prime Minister of Malaysia has made a promise to the international community that Malaysia will be reducing CO_2 emission per capita by 40% by the year 2020. This was announced at the 2009 United Nations Climate Change Conference in Copenhagen (COP-15) (Zaid et al., 2014). When EE measures are correctly implemented, Scope 2 of ISO 14064 GHG reduction related to 'purchase electricity' will go down, and gradually, the demand to build new power generation plants will be less (P9). Another direct impact would be on the depletion of Malaysia's natural resources and replanting of trees could be promoted further through CSR's activities.

Sub-Theme #11a – Energy Security and Mitigation of GHG and CO_2. The energy security achieved and the mitigation of GHC and CO_2 by CIECs of Klang Valley clients through the implementation of EE measures as stated by the research participants are as follows:

P1: "Maintain their cost of product even when the price of electricity increases."

P4: "Being cost-effective translates to reduced operating costs... It also means reducing the amount of energy being used, thus reducing the emission of the CO_2."

P5: "Implementing EE measures which will enable them to maintain their cost of production even as price of energy increases."

P9: "Carry out an energy audit, come up with a 5-year energy reduction roadmap with the 1st year focusing on the operational low-hanging fruit and use some of the savings to plow back into the subsequent year for further EE investment."

Sub-Theme #11b – Local CO_2 and GHG Reduction and Corporate Social Responsibility (CSR). The local CO_2 and GHG reduction and corporate social responsibility (CSR) by CIECs of Klang Valley clients through the implementation of EE measures as asserted by the research participants are as follows.

P1: "Implementation of EE measures indirectly lower down their CO_2 and GHG emission... Use of natural resources such as coal and fossil fuel to generate power that will be indirectly reduce, thus reducing CO_2 and GHG."

P4: "Reduce amount of CO_2 and GHG emission by having EE technologies directly translates to less fuel being burned to produce energy... Shall be on the reduced amount on the depletion of our natural resources and replanting of trees."

P8: "Only applicable to big corporate organizations; for small and medium industries, it's not significant... Power demand for power generation plant is reduced the use of natural resource such as coal and fossil fuel will also reduce."

P9: "EE measures are correctly implemented, Scope 2 of ISO 14064 GHG reduction related to 'purchased electricity' will go down, and gradually, the demand from the power generation plants will reduced and will lead to a reduction in the use of natural resources such as coal and fossil fuel to generate power will be reduce thus reducing the emission of CO_2 and GHG."

SUMMARY OF THEMES AND MAJOR FINDINGS

According to the research findings, there is a lack of strict enforcement in the policies and regulations provided by the Malaysian government (Theme #1, Sub-theme #1a, Sub-theme #1b). The major environmental policies incorporated in regard to reducing GHG and CO_2 are CSR policies, which are highly developed in larger multinational companies. Smaller or local companies, though, do not follow any environmental policies (Theme #2, Sub-theme #2a, Sub-theme #2b).

Areas pertaining to economic, finance, and marketing are EE projects with a quick payback period, or those having high returns on investment (ROI), or low investment cost, attain financial recognition by CIECs. The only financial risk with the implementation of these EE projects is the delivery of the expected return on investment (ROI) or payback (Theme #3, Sub-theme #3a, Sub-theme #3b).

There were not many finances available for EE training in in-house personnel, but the Human Resource Development Fund (HRDF) provides limited general purpose training and is presently available through the Green Technology Financing Scheme (GTFS). Furthermore, these finances are mostly affordable to the large multinational companies, which are HRDF registered, and to organizations that consume more than three million kWh of energy (Theme #4, Sub-theme #4a, Sub-theme #4b, Sub-theme #4c).

Regarding behavioral, information, and technical concerns, top managers of an organization are fully aware, whereas smaller and local organizations lack awareness regarding EE implementation. When comparing the level of awareness about EE among employees of an organization, employees of larger organizations are more aware, whereas local and SMEs organizations lack awareness of EE implementation (Theme #5, Sub-theme #5a, Sub-theme #5b, Sub-theme #5c, Sub-theme #5d).

The common type of energy management systems (EnMS) implemented and integrated by CIECs in Klang Valley is ISO 50001. Larger multinational companies and EnMS utilizing AEMAS and UNIDO guidelines mostly implement this. However, recommendations are rarely applied. Only larger or multinational companies integrate energy management systems, and smaller organizations do not integrate these systems because the cost of implementing these systems is very high (Theme #6, Sub-theme #6a, Sub-theme #6b, Sub-theme #6c, Sub-theme #6d).

According to the different methods of implementing energy-efficient technologies in the Klang Valley, the chief option is collaborating with other organizations, particularly private owned organizations that provide low-lying energy solutions. There is, though, no support from the government as well as from non-government entities (Theme #7, Sub-theme #7a, Sub-theme #7b). As for the implementation of Clean Development Mechanism (CDM), there are no funds available in the Carbon Trust, and the CDM has completely fizzled out. In reference to implementing of best available technologies (BATs), energy saving lighting systems, like the LED lighting system, were highly applied. Other techniques like low energy consuming motors and high-efficiency motors were implemented (Theme #8, Sub-theme #8a, Sub-theme #8b). As for an energy management system (EnMS), larger organizations implement ISO 50001, and smaller organization do not follow any EnMS. The facilities provided to the Klang Valley clients are certified primarily through Leadership in Energy and Environmental Design (LEED) and the Green Building Index (GBI). Equipment generally used to monitor these facilities is through an online real-time computer-based monitoring system through Building Automation System (BAS), Building Intelligent System (BIS), and Building Energy Management System (BEMS). They are largely practiced by larger organizations, and smaller organizations monitor only through human observations and manual data recording (Theme #9, Sub-theme #9a, Sub-theme #9b, Sub-theme #9c).

With regard to the final research question to find the usefulness of EE technology in terms of economic gains, it will be achieved largely through lower cost of production, which will increase the overall revenue for the

CIECs of Klang Valley (Theme #10, Sub-theme #10a, Sub-theme #10b). The cost-effective measures through the implementation of EE technologies help to achieve energy security and improve business productivity. This, in turn, will help to mitigate GHG and CO_2 and to maintain the cost of their product even when the price of electricity or business cost increases (Theme #11, Sub-theme #11a). The implementation of EE technologies will make CIECs of Klang Valley socially responsible (CSR) as EE is employed. This will in effect indirectly reduce CO_2 and GHG emission (Theme #11, Theme #11b). When energy requirement is reduced due to EE implementation, the current power plants serving the CIECs in Klang Valley will require generating less energy, which in turn will consume less natural resources like coal and fossil fuel. This will result in directly reducing GHG and CO_2 emission. The implementation of EE will create an unused capacity of the current power plants that in the future could cater for new economic growth, thus reducing the need to build new power plants.

SUMMARY

Chapter four covered the key findings and themes that emerged from the interviews. This chapter contained a discussion of the findings of the study, including the participants' characteristic data and the information obtained during individual interviews. The purpose of this phenomenological study was to investigate and bring together the lived experiences of respondents participating in this research so as to develop a deeper understanding of the attitudes of CIEC in Klang Valley through the perceptions of energy experts. Themes together with sub-themes one to five proved to be highly relevant in addressing research question RQ#1; themes together with sub-themes six to nine proved to be highly relevant in addressing research question RQ#2; and themes together with sub-themes ten and eleven proved to be highly relevant in addressing research question RQ#3. Chapter five presents discussions of the results, provides key conclusions, and provides recommendations for further research based on these findings.

Chapter 5
Conclusion and Findings

This study explored and attempted to understand the perception of energy experts on commercial and industrial electricity consumers (CIEC) of Klang Valley, Malaysia toward the adoption of energy efficiency technologies. The prior chapters contain an overview in which it provided the rationale for this study, a literature review and an analysis of the literature of previous research conducted, presentation of the methodology utilized, data collection and analysis, and other aspects related to this study.

A phenomenological approach was utilized for this study in order to explore the individual experiences of the participants, who provided energy services to CIEC in Klang Valley. This final chapter contains a discussion of the results of this study and provides key conclusions drawn from the findings. In addition, this chapter provides recommendations for further research based on these findings.

SUMMARY OF THE RESULTS

Three research questions were designed for this research study. The first was to provide evidence on the resistance of the CIECs of Klang Valley toward adopting EE postures. The second question looked into the implementation of EE technologies in Klang Valley, and the final question addressed the usefulness of EE technologies. Based on these research questions, a literature review was carried out focusing on various EE topics pertaining to policy and regulatory, economics, finance, marketing, behavior, information, technology, and associated barriers. The literature review was further conducted based on previous research carried out on the usefulness of energy-efficient technologies by other researchers, which was related to economic gains, energy cost reduction, achieving energy security, and the contribution toward CO_2 and GHG emission reduction.

Pertaining to the first research question and according to the research findings, the absence of strict enforcement in the policies and regulations by the relevant authorities of the government of Malaysia, who are responsible for EE matters, is evident in the research findings (Theme #1, Sub-theme #1a, Sub-theme #1b). Eight percent of the research participants agreed that there is a lack of enforcement by government agencies responsible for energy regulations. The major environmental policies incorporated in regard to reducing GHG and CO_2 emission are CSR policies which are highly developed in larger multinational companies, but smaller or local companies do not follow any environmental policies as confirmed by 70% of the research participants in the study's findings (Theme #2, Sub-theme #2a, Sub-theme #2b).

In areas pertaining to economic, finance, and marketing, EE projects with short payback periods, or those having high returns on investment (ROI), or low investment cost, attain financial recognition by CIECs of Klang Valley. The only financial risk that could occur when implementing these EE projects is the expected return on investment (ROI), or that payback is not met. This was affirmed by 70% of research participants in the research findings (Theme #3, Sub-theme #3a, Sub-theme #3b).

There are insufficient financial funds for EE training for in-house personnel, but the Human Resource Development Fund (HRDF) provides general purpose training, which is limited yet is presently available through the Green Technology Financing Scheme (GTFS). Furthermore, these finances are mostly affordable to the large multinational companies and organizations that consume more than three million kWh of energy biannually, which are HRDF registered and have been recognized as high energy consumer by Suruhanjaya Tenaga Malaysia (Energy Commission of Malaysia). The research findings (Theme #4, Sub-theme #4a, Sub-theme #4b) recognized that SMEs and small local organizations cannot afford to train their in-house personnel in EE due to lack of internal funding and inability to receive any external finance as they are not HRDF registered as confirmed by 90% of the research participants.

Regarding behavior, information, and technical concerns, top managers of organizations are fully aware, whereas the managers of smaller and local organizations lack awareness regarding EE implementation. This was confirmed by 70% of the research participants in the research findings (Theme #5, Sub-theme #5a, Sub-theme #5b). When comparing the level of awareness in EE among employees of an organization, employees of larger organizations are more aware, whereas the employees in local and SMEs organizations lack awareness of EE implementation as confirmed by 60% of the participants in the research findings (Theme #5, Sub-theme #5c, Sub-theme #5d).

Based on the second research question, the common type of energy management systems (EnMS) implemented and integrated into the overall management system by CIECs in Klang Valley is the ISO 50001, which is implemented mostly by larger multinational companies, and EnMS utilizing AEMAS and UNIDO guidelines and recommendations are rarely implemented. Only larger or multinational companies integrate energy management systems, and smaller organizations do not integrate these systems since the cost of implementing and managing these systems is very high. This has been substantiated by 90% of the research participants in the study's findings (Theme #6, Sub-theme #6a, Sub-theme #6b, Sub-theme #6c, Sub-theme #6d).

The second research question also revealed that there are different methods for implementing energy-efficient technologies in Klang Valley, but the best option is to collaborate with other organizations, particularly private owned ESCOs (Energy Services Companies) which provide low-lying energy solutions. At this present time, there is no support from the government as well as from non-government organizations as confirmed by 60% of research participants in the study's findings (Theme #7, Sub-theme #7a, Sub-theme #7b). As for the implementation of clean development mechanisms (CDM), there are no funds available in the Carbon Trust (Carbon Trust, 2012), and CDM has completely fizzled out (Kill, Ozinga, Pavett & Wainwright, 2010). The absence of CDM has been further confirmed by 100% of the research participants in the research findings (Theme #8, Sub-theme #8a). In reference to the implementation of best available technologies (BATs), energy saving lighting systems like LED lighting were highly implemented as based on 90% of the research participants in the research findings (Theme #8; Sub-theme #8b). Other techniques like low energy consumption and high-efficiency motors were implemented but at a lower scale. As for energy management systems (EnMS), the finding of this study concluded that larger organizations implement ISO 50001, and smaller organization do not follow any EnMS. This was ascertained by all of the research participants. The CIEC of the Klang Valley facility are certified primarily through Leadership in Energy and Environmental Design (LEED) and Green Building Index (GBI) as confirmed by 90% of the research participants in the research findings (Theme #9, Sub-theme #9a). CIECs facility equipment are generally monitored through an online real-time computer-based monitoring system through Building Automation System (BAS), Building Intelligent System (BIS), and Building Energy Management System (BEMS). These are largely practiced by larger organizations, and smaller organizations monitor only through human observations and manual data recording as was confirmed by all the research participants in the research findings (Theme #9, Sub-theme #9b, Sub-theme #9c).

With regard to the final research question to find the usefulness of EE technology in terms of economic gains, it will be achieved largely through EE implementation. This will lower the cost of production and in turn will increase the overall revenue of the CIECs of Klang Valley (Theme #10, Sub-theme #10a, Sub-theme #10b). The cost-effective measures through the implementation of EE technologies help to achieve energy security and improve business productivity. This, in turn, will help to mitigate GHG and CO_2 and to maintain the cost of their product even when the price of electricity or business cost increases (Theme #11, Sub-theme #11a). The implementation of EE measures will make CIECs of Klang Valley socially responsible as the implementation of EE technologies indirectly reduces CO_2 and GHG emissions (Theme #11, Sub-theme #11b). When energy requirement is reduced, due to EE measures, the present power plants serving the CIECs of Klang Valley will be required to generate less energy and that, in turn, will consume lesser natural resources such as coal and fossil fuel, resulting in directly reducing GHG and CO_2 emissions. The implementation of EE technologies will create an unused capacity of present power plants that in the future could cater to economic growth and thus reduce the need to build new power plants. It was unanimously affirmed by all research participants that EE implementation would bring about economic gains and energy cost reduction (Theme #10, Sub-theme #10a, Sub-theme #10b).

CIEC of Klang Valley clients can contribute to further reduction of local CO_2 and GHG emissions and can be socially responsible (CSR) by reducing energy consumption through the implementation of EE measures, which indirectly reduces the CO_2 and GHG emissions as well as the demand for power generation. When the demand for power generation reduces, natural resources such as coal and fossil fuel to generate power will eventually reduce as each of kWh reduction in electricity aids in reducing CO_2 emission by 0.694 kg. This will directly reduce CO_2 and GHG emissions and support corporate social responsibility as confirmed by 100% of the research participants in the research finding (Theme #11, Sub-theme #11a, Sub-theme #11b).

DISCUSSION OF THE RESULTS

This study, therefore, aims to contribute further understanding relating to the key issues specified above through a qualitative phenomenology investigation of the perceptions of energy experts on commercial and industrial electricity consumers of Klang Valley, Malaysia.

Chapter four presented the research participants' perception of commercial and industrial electricity consumers (CIEC) of Klang Valley, Malaysia toward the adoption of energy-efficient technologies. The impacts were outlined and summarized under eleven main themes and twenty-eight sub-themes which provided deeper insight into the participants' perceptions and experiences, which were concluded from the analysis of the interview data.

RQ#1: Resistant to Adopting Energy Efficiency Postures?

Theme #1: Policy and Regulatory: There is Insufficient Support in Terms of Meeting Policy and Regulations Requirements, and the Policies and Regulations Implemented Are Ineffective. With regards to policy and regulations, the support provided in meeting requirements are inadequate, and the policies and regulation implemented are ineffective. The Energy Commission of Malaysia has brought in main policies and regulations that come under the purview of EE, but it is not being enforced strictly (Sub-theme #1a). The Energy Commission set the regulations; however, it takes a concerted effort from all layers, including consumers, associations, NGOs, and other related parties, to work together, which is needed to make these policies and regulation effective. Usually, these policymakers provide a lot of input over the internet and web pages, but all these are rarely used by the consumers. The information regarding these implementations of policies are present on the website and in published documents if they seek it, but generally, the CIECs do not seek this information (Sub-theme #1b). Although there are procedures provided by the government to aid organizations in implementing the government policy and regulations on energy efficiency, the local government agencies that are responsible for this are lacking behind in ensuring that these procedures are utilized effectively in aiding the organizations due to a lack of enforcement (DOE, 2015).

From the secondary analysis, Kadam (2014) stated the reason for improper enforcement is due to the lack of awareness and training regarding energy management, and because manufacturing organizations are unable to formulate strong EE policies due to the manufacturing environment, where production is given more importance than energy management. A report by IEA (2010) stated that the reason for lack of enforcement might be due to lack of knowledge and capacity in incorporating EE measures among the local government, such as in reducing or managing the losses in transmission and distribution of electrical power to consumers. To overcome these issues, the government, when designing policies that relate to incentives, must take into careful consideration private and public sectors' interests since the policies cover both public (reduction of greenhouse gases) and private (energy cost savings) benefits. Governmental energy management programs are crucial in overcoming the general barriers, typically the lack of awareness on the implementation of an EnMS. Governments must provide support and guidance throughout the implementation process as recommended by Kelley, Goldberg, Magdon-Ismail, Mertsalov, and Wallace (2011) in their studies. This concludes with sub-theme #1a.

Theme #2: Implementation of Environmental Policies in Regard to the Reduction of GHG or CO_2 Has Been Neglected. On the subject of environmental policies, the economies industrialize the dependency on more sophisticated infrastructure, causing an increase in technology systems that make energy more important. The world's continued industrialization and economic growth could be potentially threatening due to the number of energy-related problems and constraints (Fawkes, Oung & Thorpe, 2016). From the research findings, there are only a few who have implemented environmental policies employing AEMAS, ISO 50001, and the ISO 14000 series standards as recommended in the guidelines. Some have implemented generic environmental policies to satisfy the organization's corporate social responsibility (CRS) policies toward the reduction of GHG and CO_2 emission. There has not been a single separate policy that is committed enough toward the environment; it is more of meeting the obligations (Sub-theme #2a). But most of these policies are being initiated only by the larger MNCs (multinational companies) by incorporating corporate social responsibility (CSR) policies toward the reduction of GHG and CO_2, but small local companies do not have strong policies toward the GHG and CO_2 reduction among the CIEC of Klang Valley since there are no specific policies or regulations requiring them to implement these initiatives and policies. The more kWh you save, the less CO_2 emissions, which means one kWh saved will reduce CO_2 emission by 0.0007 tons (Sub-theme #2b).

From the secondary data analysis for environmental policies, it was seen that most studies show that European Union countries are often responsible for the biggest part in energy consumption, especially targeting buildings, which consumes over 40 percent of total energy. It was found that an artificial intelligence model such as neural networks and support vector machines, where these technologies have a high potential for mapping environmental conditions in real situations, could be used to control energy consumption. The United Nations Environment Programme (2009), which reported on climate change and building, highlighted six major key messages. First, the building sector has the most potential of delivering significant and cost-effective GHG emission reduction. Second, countries that have not implemented EE technologies will not meet the targets on emission reduction. Third, proven policies, technologies, and knowledge are available to reduce GHG emissions. Fourth, the building industry is committed to taking action, and many countries are already playing leading roles in GHG emission reduction. Fifth, a key factor is that there are significant co-benefits, including employment, that will be created by policies encouraging EE and low GHG emission building activities. Finally, if the building sector fails in encouraging low carbon emission and EE in new building construction or in retrofitted buildings, it will lock countries into a disadvantage of poor performance buildings for decades (UNEP, 2009).

Theme #3: Financial Recognition for EE Projects and Financial Risk Faced in EE Implementation. In regard to financial concerns, the type of EE projects that get financial recognition among CIECs of Klang Valley are projects with a rapid payback period, high ROI, low investment costs, and offer energy performance contracting (EPC) (Sub-theme #3b). From a report published by EPA by the EPA in 2017, it was stated that there is also a lack of recognition from financial reports on EE technologies' beneficial contributions to the overall environment. It is perceived that the implementation of energy efficiency projects might affect the operation of CIEC's core business processes, which is one of the main concerns possibly compromising productivity (EPA, 2017).

Once a company successfully implements an EnMS, the information it provides will, in turn, address a variety of other common barriers to EE such as the perceived technical risks and financial viability of EE projects. The EE market will have to correlate with finances and the perceived technical risks that are likely to occur when implementing EE projects. The experience of the market in accepting an energy management system associates with government-led programs to stimulate and encourage companies to implement them (Kelley, Goldberg, Magdon-Ismail, Mertsalov, &, Wallace, 2011). The following are various financial risks identified in this study that are being faced by the clients of CIEC of Klang Valley in EE projects implementation, including the non-deliverance of the expected savings within the stipulated time period; thus, the return on investment is not being met, leading to financial risk. When tariffs are being revised, the calculations on the payback and ROI will differ, leading to financial risks. Financial risk is not foreseen if the project is implemented after a thorough energy audit has been carried out (Sub-theme #3a). Furthermore, Backman (2017) stated that an energy audit does not affect the investment plans directly, and investment plans and energy audit should be treated separately. Lack of technological information and funds are major barriers in overcoming municipal energy policies, and these are the main barriers in increasing EE measures due to the lack of information in various areas. Some of the information lacking includes the real cost of implementing new EE technologies, the total cost of consumer's own energy consumption, knowledge on EE technologies, and the cost of operating and maintaining EE equipment. If these costs are not taken into consideration, the particular EE project will become a high-risk investment if the aim is to have high returns. These non-transparencies of information can be misleading to an organization, especially in conducting the feasibility of investing in an EE project (European Commission, 2007).

Theme #4: Availability of Finances for EE Training and the Consumers Who Are Able to Afford In-House Energy Personnel. According to Kadam (2012), the non-prioritization of finances for EE training for personnel has led to many missed opportunities that could have been brought by EE implementation (Kadam, 2014). The types of financial funds available for the CIEC of Klang Valley toward EE training for their in-house personnel are insufficient. However, in the past, there were some funds available under the Malaysian Industry Development Agency (MIDA) worth about RM200,000 to train the members of the Malaysian Energy Professional Association (MEPA) and many others from the institution (MIDA, 2017). Unfortunately, the funding has since ceased (Sub-theme #4a). Many other fundings are jointly available from UNIDO, which was a five-year program that just ended in 2016. The program provided free training for people from the industrial sector that showed evidence of commitment from the authorities. Some of the external funding available was from ESCOs that provided some financing for EE training. There were limited funds available from the Malaysian government agencies for general purposes only. Some funding was also made available by NGOs and from the Malaysian government's Human Resource Development Fund (HRDF). However, the 100 percent refund through HRDF is still available through the Green Technology Financing Scheme (GTFS) (Sub-theme #4b).

According to UNIDO (2017), the availability of skilled and knowledgeable staff in energy management is limited due to affordability in small and medium industries. Skills required to carry out EE projects in these industries could be outsourced to external energy management experts to implement their in-house EE projects. This will help overcome barriers like the lack of in-house energy personnel. Therefore, according to the research findings, the types of clients who could afford in-house energy personnel are mostly large MNCs or big energy consumers and not necessarily MNCs. National companies that consume more than three million kWh (kilo-Watts-hour) in any period of six months are the ones who will definitely see the benefit of having in-house energy personnel. Consumers who have exceeded the 500 thousand kWh in any six months period engage the services of Registered Electrical Energy Managers (REEM) personnel due to regulatory requirements (Sub-theme #4c). A report by DOE (2015) stated that there is a lack of available knowledge within federal, state, and utility incentives for end-use EE measures, which can lead to missed opportunities. Another opportunity missed by consumers due to lack of insufficient in-house technical expert and non-participation is demand response programs. Demand response is a program where consumers of electricity can take financial advantage by changing their electric usage from their normal consumption patterns in response to changes in the price of electricity over time, or to provide incentive payments designed to induce lower electricity use at times of high wholesale market prices or when system reliability is jeopardized.

With regards to the discussion on behavior, information, and technology, engaging installation personnel who are inadequately trained in implementing EE technologies will influence the behavior of consumers to have confidence in the project's success (UNIDO, 2017).

Theme #5: Awareness of Energy Management Techniques among Top Managers and Employees Needs to be Increased. From the analysis of data by Kadam (2014), there is a lack of awareness in applying energy management techniques and insufficient training programs in energy management. The findings from this study reveal that in large multinational companies, members of the top management are fully aware of energy efficiency (EE) and priority is only given to EE as part of its corporate social responsibility (CSR) (Sub-theme #5b). Whereas in local, smaller, and medium enterprises (SMEs), there is a lack of concern for EE even though top management is aware of EE (Sub-theme #5a).

The research findings also concluded that awareness among employees regarding EE implementation is higher among larger multinational companies since they conduct regular, internal energy awareness programs within their organizations to make employees conscious of the importance of EE (Sub-theme #5d). To encourage the involvement of employees in energy consciousness, energy awareness campaigns are to hold competitions between different departments within an organization to determine the most effective energy-efficient solutions and to find the solutions that provide the most innovative ideas for energy conservation. However, these kinds of campaigns and competitions will cost money, and most of the smaller and local companies cannot afford these kinds of campaigns in order to promote employee awareness (Sub-theme #5c). A report by the Department of Energy (2015) stated that employees are the major barriers to the adaptation of EE technologies. From a report by EDF Climate Corps, the companies that were surveyed had little or no knowledge on how EE technology affects their performance in terms of energy use. An example of this is the lack of information and incomplete standardization related to the implementation of combined heat and power (CHP) (DOE, 2015).

Regardless of many guidelines and policies with respect to energy efficiency, most top management is not aware of their responsibilities toward the implementation of EE projects. Low awareness of the benefits of EE within the top management of companies will be reflected in the behavior of their own employees toward energy savings (Apeaning & Thollander, 2013). Energy management requires staff within an organization to acquire trained skills and knowledge in EE technologies and projects, which is lacking in many small and medium industries (UNIDO, 2017). According to the research findings, members of top management in large multinational companies are fully aware of energy efficiency (EE) and priority is given to EE only when the company wants to boost its branding or image, for example, to promote its name via corporate social responsibility (CSR) activities (Sub-theme #5b). Whereas in local, smaller and medium enterprises (SMEs), there is a lack of concern for EE even though the top management is aware of EE, but due to tight profit margins and high competition among SMEs of a particular industry, the top management of SMEs tend to focus on gaining profit by focusing on the business revenue rather than implementing EE. Furthermore, SMEs are unable to afford the capital cost of implementing EE measures (Sub-theme #5a).

RQ#2: How Can Energy Efficiency Technologies be Implemented?

Theme #6: CIECs Who Have Standard EnMS Implemented and Integrate It Into the Overall Management System. From the research findings, the standard energy management systems implemented by the CIEC of Klang Valley is ISO 50001, and it is a voluntary international standard for energy management system (EnMS). The International Organization for Standardization in collaboration with the United Nation Industrial Development Organization (UNIDO) developed the ISO 50001 in response to climate change and the need to have a standard global energy management system. The standard was developed to equip organizations with the requirement of an EnMS that derives its standard from national and regional energy management standards, specifications, and regulations. The key focus of EnMS is that it involves all levels and functions of the company and requires the participation of top management. Therefore, from the findings of this study, it was found that mostly large MNCs have implemented ISO 50001 energy management systems as it ensures the continuous motivation of employees, which is a pertinent element in the effective functioning of the EnMS framework (Sub-theme #6b), and it is incorporated into the existing overall management system of an organization (Sub-theme #6d). The implementation of ISO 50001 is an effective means of overcoming the prevalent informational, institutional, and behavioral barriers to EE implementation (OECD, 2015).

Therefore, it was identified from the research findings that the integration of EnMS into an overall management system is implemented mostly in large organizations and large MNCs (Sub-theme # #6a). Even though it is costly, it is incorporated into the overall organization in MNCs, whereas it is not implemented in the small organizations due to non-affordability, and hence, it is not integrated into the smaller organization's overall management system (Sub-theme #6c). However, a report by OECD (2015) stated that in the promotion of EnMS uptake, there is a range of possible incentives due to the large variation in the market and regulatory forces in different countries. EnMS incentives can improve the participation of industries in implementing EE; however, it can also inadvertently lead to restricting private sector investment support. There are many possible hazards that can happen if incentives are unnecessarily used or are not well adapted to the real need of the industry. The main barrier to EnMS uptake is the lack of knowledge rather than financial constraints (OECD, 2015). Huang (2011) stated that the implementation of an energy management standard within an organization requires changes in existing institutional practices toward energy, a process that could benefit from technical assistance from experts outside the organization. The benefit is an organization's staff's familiarity with management systems like quality, safety, and environment; understanding of the dynamics of establishing a management system; and its successful integration into the organization's corporate culture.

Theme #7: Organizations Offering Collaboration for EE Implementation Are Limited. CIEC of Klang Valley collaborates with private organizations since CIECs are not receiving any collaboration or support from the government or NGOs for implementing EE (Sub-theme #7a). Certain private organizations like Energy Service Companies (ESCOs) and Malaysian Energy Professional Association (MEPA) offer low-cost energy services and solutions by helping in implementing EE action plans and services which are limited due to the high cost to implement EE projects based on the EPC model as these ESCOs are unable to borrow from financial institutes that perceive EE projects as high-risk (Sub-theme #7b). The European Commission (2016) suggested that organizations should carry out continual improvement on their EE programs where their objectives are to attain long-term improvements with the performance measured periodical. Finally, a study by UNEP (2015) stated that there is lack of implementation of policies of EE programs formulated by government agencies and their local Department of Energy (DOE) (DOE, 2015). Furthermore, a study from IPEEC (2016) identified that a practical Energy Efficiency Action Plan (EEAP) would strengthen voluntary EE collaboration in a flexible way as was adopted by the group of 20 nations (G20) members in 2014 based on the activities that best reflected their domestic priorities and interests. It permits countries through an opt-in basis to share their knowledge, experiences, and resources. The Energy Efficiency Leading Programme (EELP), apart from covering the existing activities under the EEAP on vehicles, networked devices, finance, buildings, industrial processes, and electricity generation, expands its work areas to include five new key areas of collaboration: Super-efficient Equipment and Appliances Deployment (SEAD) initiative, Best Available Technologies and Practices (TOP TENs), District Energy Systems (DES), Energy Efficiency Knowledge Sharing Framework, and Energy End-Use-Data and Energy Efficiency Metrics.

Theme #8: Implementation of CDM and BATs. Eang (2015) identified that the Clean Development Mechanism (UNFCCC, 1998) process can be implemented to support EE programs for buildings and GHG emissions data can be developed in maintaining baselines for a consistent approach on reporting and monitoring of performance. But this is in contrast to the research finding where the research participants asserted that the CIEC of Klang Valley no longer enjoy the benefits of the CDM program as it is no longer available (Sub-theme #8a). Under the Clean Development Mechanism (CDM) (UNFCCC, 2006), emission-reduction projects in developing countries can earn certified emission reduction credits. The Kyoto Protocol (UNFCCC, 1998) allows for these saleable credits to be used by industrialized countries to meet a part of their emission reduction targets. Governments then issue permits up to the agreed limit, and these are either given free or auctioned to companies in the sector. They work by setting an overall limit or cap on the amount of emissions that are allowed from significant sources of carbon, including the power industry, automotive and air travel, at regional, national, and international levels. If a company curbs its own carbon significantly, it can trade the excess permits on the carbon market for cash. If it is not able to limit its emissions, the company may have to buy extra permits. Therefore, this carbon credit could be traded, sold, and utilized by industrialized nations to purchase the carbon credit to pay for their excess CO_2 emissions under the Kyoto Protocol requirement. Creating a market in something with no intrinsic value, such as carbon dioxide, is very difficult. There is a need to promote scarcity, which should strictly limit the right to emit CO_2 so that it can be traded. A glut of permits was created due to this. These permits have often been given away for free, which has led to a collapse in the price, and no effective reduction in emissions compounded, allowing offset permits gained from paying for pollution reduction in poorer countries to be allowed to be traded as well (Kill, Ozinga, Pavett & Wainwright, 2010). Due to the above-mentioned issues, the CDM process has died off in Klang Valley, and a mechanism to gauge CO_2 reduction is no longer available.

The BATs (best available technologies) that can commonly be found among the CIEC of Klang Valley are LED lighting system, low energy consuming motors, VFDs (Variable Frequency Drives) for motors, intelligent chiller control, power optimize, inverter type air-conditioning, air-water system, and high-efficiency motors. But the most adopted BAT is the LED, which is very straight forward. The implementation time for LED is very short and is completely manageable cost-wise (Sub-theme #7b). A report by McKane, Scheihing & Williams (2007) and Seai (2012) stated that EE in industrial sectors is usually achieved by changing the operation of an industrial site utilizing BATs. The first step in identifying and prioritizing the full range of EE opportunity is to implement an energy management system in advance of other major investment decisions related to physical capital, while stand-alone BAT investments are still important in achieving the full potential of EE. It has been shown that experienced industrial companies can save up to 10–30% of their annual energy consumption and a similar reduction in their operating costs through better energy management involving operational changes alone.

Theme #9: There Is a Lack of EE Certification on Facilities, Facility Equipment Monitoring, and Online Execution Systems. According to findings of this study, the CIEC of Klang Valley facilities that have been recognized as energy efficient are only a few and have been certified through the LEED (Leadership in Energy and Environmental Design) and GBI (Green Building Index) certification standards (Sub-theme #9a). The local certification is through the Penarafan Hijau (PH – Green Certification), GreenPASS, and Malaysian Carbon Reduction Assessment Tool (MyCrest) (P4, 2017), whereas international recognition comes from American Leadership in Energy and Environment Design (LEED) and Greenmarks (Singapore). The importance of facilities being certified as energy efficient has been addressed by the United Nations Environment Programme (UNEP) (2009), which reported on climate change and building, where six major key messages were highlighted. One of the six major key messages was that in order to prove that the GHG reduction goals are supported, buildings that have implemented a reduction in GHG emission have to be certified by Nationally Appropriate Mitigating Action (NAMA).

The Department of Energy (2015) stated that the energy management system with its real-time data acquisition system and with the appropriate analytical tools can optimize energy usage of energy consuming equipment in the manufacturing facilities and provide established statistical reports on the performance of the organization's EE initiatives. These statistical reports can provide valuable information to further improve the existing EE projects and to promote new measures. In most cases, the above-mentioned energy management systems are lacking in the manufacturing sectors (DOE, 2015). From the research findings, the facility equipment of the CIECs of Klang Valley are generally monitored through an online real-time computerized monitoring system, which is through field gauges, meters, and sensors. Other data acquisition devices are carried out by utilizing Building Automation System (BAS), Building Intelligent System (BIS), and Building Energy Management System (BEMS). These systems are generally being utilized by large-scale industries, whereas human observations, manual recordings, and checklists are generally utilized by small-scale industries (Sub-theme #9b) due to affordability.

The types of online execution systems that are available to the end-user via user interfaces, which help manage, monitor, and control energy usages which are utilized by the CIEC of Klang Valley, are Building Automation System (BAS), Building Intelligent System (BIS), and Building Energy Management System (BEMS); they are integrated with customized software for a specific engineering system. These systems are usually utilized by larger organizations due to it being expensive. However, the small-scale organizations utilize manual logging and assessment on a regular basis due to affordability (Sub-theme #9c). A study by Tanaka, Watanabe & Endou (2010) stated that the users of conventional energy management systems are supposed to be utilized only by the managers of energy supplying utility equipment. However, the Enerize E3, a factory energy management system developed by Yokogawa, Japan, is the first system in the industry to standardize energy management, encouraging all members in a factory to participate in energy saving activities. It assumes a wider variety of people as its users, and hence, such a model system might act as a basis for implementing energy management systems in the CIECs of Klang Valley, Malaysia.

RQ#3: How Do Consumers Find Energy-Efficient Technology Useful?

Theme #10: EE Implementation Provides Economic Gains and Energy Cost Reduction. From the research findings, the CIECs of Klang Valley could benefit in terms of economic gains and optimized energy cost while expanding their business, which generally requires more energy consumption. The CIECs of Klang Valley could be cost-effective in achieving energy security in improving business productivity and to mitigate GHG and CO_2. This could contribute to the reduction of local CO_2 and GHG emissions and are socially responsible (CSR) if energy efficiency technologies are adopted.

It is shown that the kind of economic gains developed include a lower cost of production that will increase CIEC profitability, which will lead them to become competitive against others in a similar industry. Moreover, through their increased competitiveness, their products can be offered at a lower price. A typical commercial building can reduce its energy consumption by 40% when compared to a baseline kWh (P9, McKane, Scheihing & Williams, 2007; Seai, 2012). The gains, even though small, cannot be explained in terms of dollars and cents; it is expressed in kWh. With business expansion, it is obvious that the absolute energy consumption will increase, but at the same time, the increase in efficiency of production due to EE will also correspondingly increase, thus resulting in an increase in revenue. This is effective when the production cost decreases and the revenue increases (Sub-theme #10a).

Pertaining to the energy cost, it is usually on the average of about 20% of the production cost. Therefore, any EE implementation will only translate to 10% to 20% energy reduction, which translates to 2% to 4% of the production cost. Hence, implementing EE technologies helps CIECs of the Klang Valley in terms of competitiveness, profitability, and achieving energy security as it reduces energy cost. Apeaning (2012) stated that economic gains related to cost reductions resulting from lowered energy use and threats of rising energy prices are the most important drivers for implementing EE measures or technologies; also, government efficiency requirements is another important promoting factor for implementation (Sub-theme #10b).

The following are a couple of projects undertaken by some of this study's participants demonstrating the benefits of adopting of EE technologies:

Project No. 1:

Project:	Fast food outlet
Location:	Klang, Valley, Malaysia
Date:	1st March to 29th August 2017
Equipment:	Split air-condition unit
No of Units:	2
Saving Method:	Utilization of advanced temperature sensing and control algorithm to identify exactly when refrigerant compressor in air-conditioning can be switched off.
Results:	The following are the results obtained:

The bar chart in Figure 19 illustrates the energy consumption trends over two periods of 15 days duration for a split unit type of air-condition identified as unit no.1. The first period was from 1st to 29th March 2017, and the second period was from 1st to 29th August 2017. The first period is represented by the red bars pointing out the electricity consumption in kWh of the above-mentioned air-condition unit before the installation of the energy efficiency device, e.g. the baseline consumption. The second period is represented by the blue bars and points out the electricity consumption in kWh after the installation of the energy efficiency device, e.g. the saving mode electricity consumption. The average baseline and savings mode electricity consumption were computed.

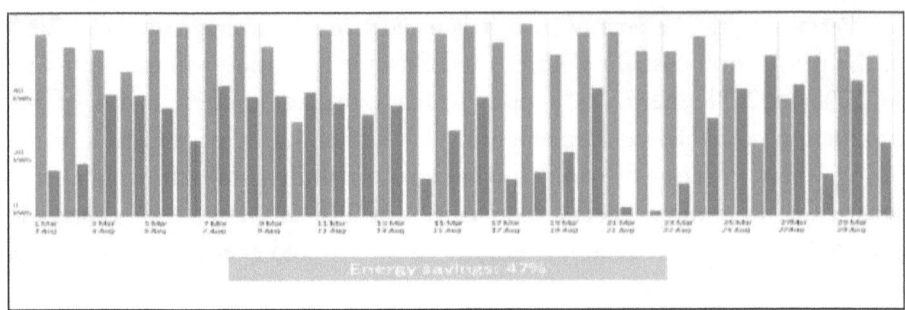

Figure 19 Saving results for air-condition No. 1. Adapted from Seido Solutions (2018).

The average savings percentage achieved by the implementation of the energy efficiency device, which utilized advanced temperature sensing and control algorithm to intelligently cut-in and cut-off the air-condition refrigerant compressor, was computed as follows:

$$Saving\ \% = \left\{ \frac{\left[\left(\sum \frac{kWhredbars}{12}\right) - \left(\sum \frac{kWhbluebars}{12}\right)\right]}{\left(\sum \frac{kWhredbars}{12}\right)} \right\} * 100\%$$

Applying the above computation method, the average savings percentage achieved for unit no.1 air-condition was 47%.

The bar chart in Figure 20 illustrates the energy consumption trends over two periods of 14 days duration for a split unit type of air-condition identified as unit no. 2. The first period was from 11th to 24th March 2017, and the second period was from 11th to 24th August 2017. The first period is represented by the red bars indicating the electricity consumption in kWh of the above-mentioned air-condition before the installation of the energy efficiency device, e.g. the baseline consumption. The second period is represented by the blue bars showing the electricity consumption in kWh after the installation of the energy efficiency device, e.g. the saving mode electricity consumption.

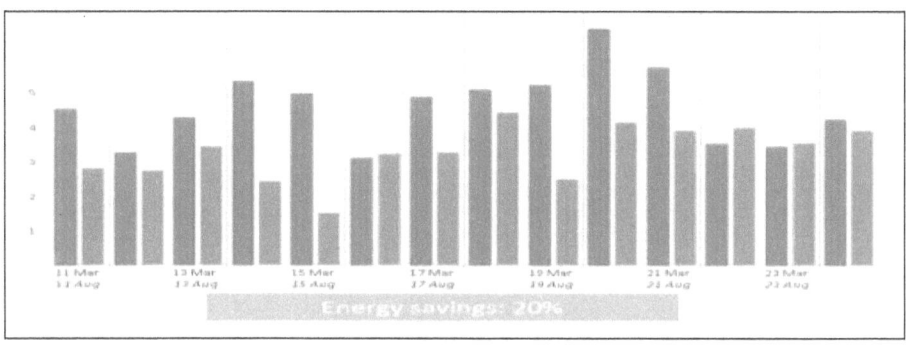

Figure 20 Savings results for air-condition No. 2. Adapted from Seido Solutions (2018).

The average baseline and savings mode electricity consumption were computed. The average savings percentage achieved by the implementation of the energy efficiency device, which utilized advance temperature sensing and control algorithm to intelligently cut-in and cut-off the air-condition refrigerant compressor, was computed as follows:

$$Saving\ \% = \left\{ \frac{\left[\left(\sum \frac{kWhredbars}{14}\right) - \left(\sum \frac{kWhbluebars}{14}\right)\right]}{\left(\sum \frac{kWhredbars}{14}\right)} \right\} * 100\%$$

Utilizing the above method of computation, the savings achieved for unit no.1 air-condition was 20%.

The bar chart in Figure 21 illustrates the energy consumption trends over two periods of 12 days duration for two split unit type of air-condition identified as units no. 1 and no. 2. The first period was from 3rd to 14th October 2017, and the second period was from 3rd to 14th November 2017. The first period is represented by the red bars showing the electricity consumption in kWh of the above-mentioned air-condition units before the installation of the energy efficiency device, e.g. the baseline consumption. The second period is represented by the blue bars illustrating the electricity consumption in kWh after the installation of the energy efficiency device, e.g. the saving mode electricity consumption. The average baseline and savings mode electricity consumption were computed.

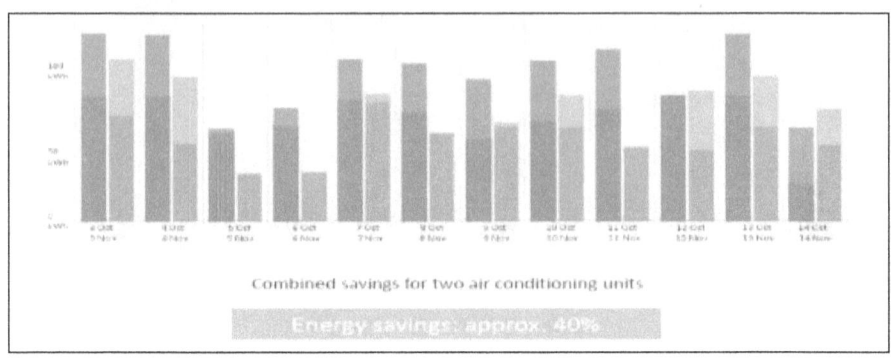

Figure 21 Combined savings results for air-condition No. 1 & No. 2. Adapted from Seido Solutions (2018).

The average savings percentage achieved by the implementation of a single energy efficiency device, which utilized advance temperature sensing and control algorithm to intelligently cut-in and cut-off the two units of air-condition refrigerant compressor, was computed as follows:

$$Saving\ \% = \left\{ \frac{\left[\left(\sum \frac{kWhredbars}{12}\right) - \left(\sum \frac{kWhbluebars}{12}\right)\right]}{\left(\sum \frac{kWhredbars}{12}\right)} \right\} * 100\%$$

Utilizing the above method of computation, the savings achieved for the combined units no. 1 and no. 2 air-condition was 40%.

Project No. 2:

Project:	Retail Pharmacy
Location:	Klang, Valley, Malaysia
Date:	30th August to 19th September 2017
Equipment:	Lighting
No of Units:	260 light units
Saving Method:	Replacement of T8 fluorescent light bulbs with Nano LED light bulbs
Results:	The following are the results obtained:

The bar chart in Figure 22 illustrates the energy consumption trends over two periods of 20 days duration for a retrofitting project involving electricity consumption improvement project for lights. The first period was from 29th July to 17th August 2017, and the second period was from 30th August to 19th September 2017. The first period is represented by the red bars illustrating the electricity consumption in kWh with the existing lighting system, which consisted of 232 units of 4 ft T8 fluorescent and 28 units of 2 ft T8 fluorescent lights, e.g. the baseline consumption. The second period is represented by the blue bars showing the electricity consumption in kWh after the one to one replacement of the existing lights to 2 ft and 4 ft Seido Nano LED lights, e.g. the saving mode electricity consumption. The average baseline and savings mode electricity consumption were computed.

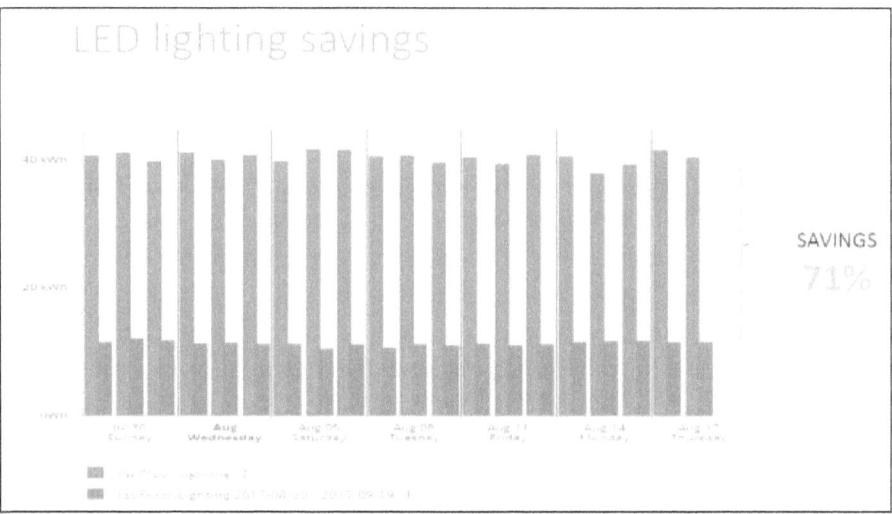

Figure 22 Savings by retrofitting to Nano LED Bulbs. Adapted from Seido Solution (2018).

Table 19 Summary Of Nano Led Lighting Retrofitting Project. Adapted From Siedo Solutions (2018)

Before	After	Qty
4ft T8 fluorescent	4ft Seido Nano LED	232
2ft T8 fluorescent	2 ft Seido Nano LED	28
	Total units of lights	260
	Potential savings per month	RM1,394.82
	Potential savings per year	RM16,737.87
	% of savings	71%

The average savings percentage achieved by retrofitting T8 fluorescent T8 to Siedo Nano LED lights was computed as follows:

$$Saving\ \% = \left\{ \frac{\left[\left(\sum \frac{kWhredbars}{20}\right) - \left(\sum \frac{kWhbluebars}{20}\right)\right]}{\left(\sum \frac{kWhredbars}{20}\right)} \right\} * 100\%$$

Utilizing the above method of computation, the savings achieved for the retrofitting project for lights was 71% and the monetary savings was RM16,737 annually (see Table 19).

Theme #11: Implementation of EE Technologies Ensures Energy Security by Mitigating GHG and CO_2 Levels to Promote Corporate Social Responsibility (CSR). The CIECs of Klang Valley have achieved energy security, improved business productivity, and mitigated GHG and CO_2 emission by implementing EE measures, which enables them to lower or maintain their product cost even when the price of energy increases. The reduction in electricity due to EE measures will compensate for the increment in energy price. This further achieves energy security by improving or maintaining business productivity and, in turn, mitigates GHG and CO_2 (Sub-theme #11a). This has been substantiated by a report by Cambridge Econometrics (2015) stating that overall benefits of implementing EE technologies are to preserve good health, create an increase in employment, increase the value of buildings or facilities, industrial competitiveness, and to increase energy security for the future. As a further measure, CIEC of Klang Valley clients should carry out an energy audit to come up with a five-year energy reduction roadmap within the 1st year by focusing on the operational low-hanging fruit by utilizing a part of their savings to reinvest into the subsequent year as EE investments. For example, a part of the previous EE saving will pay for the future EE project investment. And finally, it can be cost-effective to achieve energy security by being conscious of the latest development in EE, establish benchmarks in specific energy consumption, and compare it against similar industry benchmarks in order to gauge the effectiveness of their energy improvement programs (Sub-theme #11a). This is affirmed by a report by Cagno, Worrell, Trianni, and Pugliese (2013), indicating that the willingness to accept and utilize new technologies are essential in spite of successful implementation. Environmental assessment is normally based on achievable targets, which can be realistic and achievable. Decision makers will also need to learn about useful tools available for further innovations and future development in energy efficiency technology. Understanding these and the barriers associated with it will help in the process of energy efficiency technologies implementation.

CIEC of Klang Valley clients can contribute to further reduction of local CO_2 and GHG emissions and can be socially responsible (CSR) by reducing electricity consumption through implementation of EE measures, which indirectly reduces the CO_2 and GHG emission as well as the demand for power generation. When the demand for power generation reduces, natural resources such as coal and fossil fuel to generate power will eventually reduce as each of kWh reduction in electricity will help to reduce CO_2 emission by 0.694 kg of CO_2 (P6). This will directly reduce CO_2 and GHG emission and support corporate social responsibility (Sub-theme #11b). The Prime Minister of Malaysia has made a promise to the international community that Malaysia will be reducing CO_2 emission per capita by 40% by the year 2020, which was announced at the 2009 United Nations Climate Change Conference in Copenhagen (COP-15) (Zaid, Myeda, Mahyuddin & Sulaiman, 2014). When EE measures are correctly implemented, Scope 2 of ISO 14064 GHG reduction related to 'purchase electricity' will go down, and gradually, the demand to build new power generation plants will be less (Sub-theme #11b). Another direct impact would be on the depletion of Malaysia's natural resources and replanting of trees could be promoted further through CSR's activities (Zaid, Myeda, Mahyuddin & Sulaiman, 2015).

CONCLUSION

The main purpose of the research is identifying the behavior of commercial and industrial electricity consumers in Klang Valley, Malaysia and their impact toward adapting to the implementation of EE technologies. The study aimed to identify the various perceptions of energy experts on the commercial and industrial consumers of electricity in regard to the adoption of EE technologies.

With regards to the policies and regulations implemented by Energy Commission of Malaysia, there is an absence of enforcement by the commission, and so, CIECs of Klang Valley are not keen to take part in EE. Also, the government has not provided adequate incentives to implement EE. There is strict regulation imposed by the government or by other responsible agencies toward EE implementation, but there is lack of enforcement because there has not been any CIEC in Klang Valley who has been fined for non-compliance to this date according to the research participants. (Theme #1, Sub-theme #1a, Sub-theme #1b, Kadam, 2014).

The financial risk faced by these clients was the non-delivery of the expected savings within the stipulated payback period and not meeting the expected ROI. Only the multinational organizations had access to the internal funds for implementing EE measures, whereas access to external funds is limited and not many financial institutions support EE projects. When financial institutions offer funding for EE projects, the interest rates are high, which then impacts ROI, and the payback period makes most EE projects financially non-feasible. The CIECs of Klang Valley generally do not give priority to EE projects unless it is mandated by regulations. Improvement of projects concerning core business processes is given higher priority over EE projects as they are familiar with it and the risk in terms of failure is less when compared to EE projects, which they are not too familiar with to implement.

Data from IEA (2011) shows that among Southeast Asian nations, Malaysia has relatively high CO_2 emission rate per capita and per GDP measures at 5.97 (tonnes CO_2 per capita) and 1.20 (tonnes CO_2 per GDP) respectively (Othman & Yahoo, 2014). The Malaysian government has failed to recognize that subsidized electricity has created electricity waste among the consumers, and this creates a disincentive for EE project implementation among the CIEC in Klang Valley. In Malaysia, electricity generation is subsidized through a centrally imposed low gas price. Petronas, a producer and distributor of gas, is required by the Malaysian government to sell electricity generators at a controlled price of MYR 10.70 (US$3) per million metric British Thermal Units (MMBTU), and due to this, Malaysia has the second lowest end-user gas prices among the ASEAN nations, behind Brunei; therefore, every kWh wasted by CIECs of Klang Valley depletes the government reserves proportionally to the amount of electricity subsidized by the government (IISD, 2013).

The best way in-house personnel can gain information on the available EE technologies is by attending courses on understanding energy management systems, but there are limited finances available to the CIEC of Klang Valley clients for EE training for their in-house personnel (Theme #4,

Sub-theme #4a, Sub-theme #4b, Sub-theme #4c). According to Backman (2017), decision makers in small organizations were found to be ill-equipped and ill-informed in making EE implementation decisions due to information barriers. The transaction cost, which is the cost of gathering, assessing, and applying information about energy savings potentials and measures, as well as the cost associated with finding and negotiating the contracts with potential suppliers, consultants, or installers, or the cost of implementing, monitoring, and enforcing contracts must be included in all EE projects apart from the cost of the technology itself. Since EE measures are ideologically and technologically complex, the cost of transaction will be above average, and as a result, information barriers will be more significant in EE technologies than in other technologies. Hence, to overcome the information barrier in EE technologies, a study done by Tremblay, Lalancette, and Roseveare (2012) concluded that the effective supply of relevant information of the right quality and the education and training of the consumer are important contributions to overcoming the barrier posed by the lack of technical capacity. Dunstan, Daly, Langham, Boronyak, and Rutovitz (2011) further confirmed this in a working paper that extensive and intensive training programs should tackle the barrier of inadequate technical and managerial skills for the implementation of energy efficiency improvements. According to Worrell et al. (2001), lack of information is likely to be even more of a barrier to EE measures in developing countries than in developed countries. First, the information infrastructure at the private and public levels tends to be less developed than in industrialized countries. For example, energy management systems or energy benchmarking are less pervasive in developing countries. Likewise, developing countries suffer from limited public capacity for information dissemination, limited private technical capacity to access information (e.g. via internet), or lack of intermediary institutions providing information on the energy use of companies, processes, or technologies (e.g. via sector associations or company networks). Second, since acquiring and processing information depends on human capital infrastructure, companies in developing countries are less suited to effectively using existing information. Also, relevant information, for example, technology performance, may only be available in a foreign language (OECD, 2011). EE policies must address first the information barrier before additional policy interventions can be successful.

Once the energy management system has been implemented in an organization, continuous audits must be carried out in order to identify the low-hanging EE projects. Upon implementing the low-hanging projects, next would be to identify areas of high energy consumption. Usually, a high energy consumption system will require an in-depth technological understanding of how the system works before the gaps can be identified. This technological training must be sorted from the original system designer, manufacturer, or from associations or institutes supporting a particular industry (Theme #4, Sub-theme #4a, Sub-theme #4b, Sub-theme #4c). The secondary data of this research confirms that if there is a lack of technical information on new EE technology products, there need to be steps implemented by authorities to ensure the public has access to the information or the technology. Most of the energy consuming practices rely on sophisticated devices that are actually based on domestic and industrial appliances. Observing and understanding demonstrations of EE technology itself will provide knowledge on how to implement these technologies as a start (EPA, 2009).

The Clean Development Mechanism (CDM) permits industrialized countries to buy certified emission reduction (CER) units or carbon credits which are traded in the emission trading scheme from CDM emission projects in developing countries to meet their emission reduction target set under the Kyoto Protocol. The calculation of emission reduction is based on the emission that would have occurred without the project minus the emission after the implementation of the project (UNFCCC, 2006). Creating a market in something with no intrinsic value, such as carbon dioxide, is very difficult. There is a need to promote scarcity, which should strictly limit the right to emit CO_2 so that it can be traded, and due to this, there is a glut of permits created. These permits have often been given away for free, which has led to a collapse in the price with no effective reductions in emissions compounded, allowing an offset of permits gained from paying for pollution reductions in poorer countries to being allowed to be traded as well (Kill, Ozinga, Pavett & Wainwright, 2010). Due to the above-mentioned issues in carbon trading, and from the primary research data, it shows that the CDM process has died off in Klang Valley (Sub-theme #8a).

The best available technologies (BATs) that can be commonly found among the CIEC of Klang Valley clients are LED lighting system, low energy consuming motors, VFDs for motors, intelligent chiller control, power optimize, inverter type air-conditioning, air-water system, and high-efficiency motors. Among all the widely used BATs, LED, which is very straight forward to implement, can be installed in a relatively short time period and has a relatively low life-cycle cost (McKane, Scheihing & Williams, 2007; Sub-theme #8b).

The energy management systems (EnMS) implemented by the CIECs of Klang Valley clients is ISO 50001, which is being implemented only by large-scale organizations. As for small-scale organizations, this is not being implemented since the cost to implement and manage is very high (Sub-theme #6b). Among the CIEC of Klang Valley facilities that have been recognized as energy efficient, the majority have been certified through the LEED (Leadership in Energy and Environmental Design) and GBI (Green Building Index) certification standards. The local certification is through the Penarafan Hijau (PH – Green Certification), GreenPASS, and Malaysian Carbon Reduction Assessment Tool (MyCrest) (P4, 2017). International recognition comes from American Leadership in Energy and Environment Design (LEED) and Greenmarks (Singapore) (Theme #6).

As per the research findings, the kind of economic gains developed is the lower cost of production, which increases CIEC of Klang Valley clients' profitability. This will lead them to become competitive against similar industrial players as they are able to offer their products at a lower price. A typical commercial building can reduce its energy consumption by 40 percent when compared to a baseline kWh (Sub-theme #10a). This has been further substantiated from secondary data that economic gains related to cost reductions resulting from lowered energy use and threats of rising energy prices are the most important drivers for implementing EE measures or technologies. In addition, government efficiency requirements are another important promoting factor for EE implementation (Apeaning R., 2012).

Conceptual Framework

The basic findings for this research study clearly show that the attitudes and behavior of the CIECs of Klang Valley developed several factors that created barriers in the adoption of EE technologies. The organization size of the CIECs of Klang Valley has direct co-relation to non-availability of finances for EE training (Sub-theme #4a), non-awareness of EE management techniques among top managers (Sub-theme #5a), non-awareness of EE management techniques among employees (Sub-themes #5c), non-implementation of EnMS (Sub-theme #6a), and non-integration of EnMS into overall management system (Theme # 6c). Additionally, the lack of governmental non-support and ineffectiveness of policies and regulations (Sub-theme #1a), high-risk in EE projects which do not get financial recognition (Sub-theme #3a), and limited organizations offering collaboration for EE implementation (Theme #7) affect the attitudes and behavior of CIEC in Klang Valley, Malaysia. These barriers are further discussed and summarized with other associated barriers below. Figure 23 illustrates the conceptual framework of this research finding.

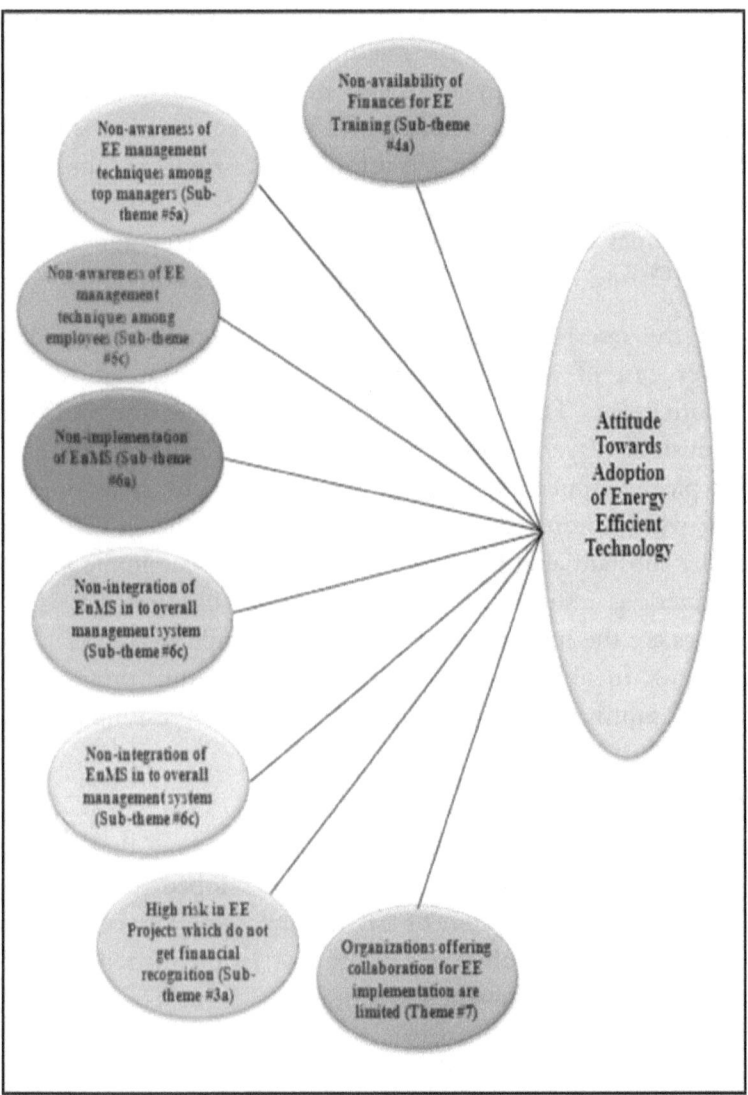

Figure 23 The Research Conceptual Framework. Source: Figure derived by author.

Non-Awareness of EE Management Techniques. The lack of awareness among CIEC in Klang Valley, as deduced by this research, was generally found among low-income organization where the top management, due to the pressures of meeting the business bottom line, only focused on improving their core business efficiency and do not give priority to EE. The employees lack motivation and use EE very little because of insufficient financial funds, training in the aspects of management, and the lack of implementation of EE. These were the barriers that restrict the adoption of EE technologies (Apeaning & Thollander, 2013; DOE, 2015; OECD, 2015; UNIDO, 2017; European Commission, 2017; Theme #5). The awareness of EE among the top management and employees of high-income CIECs of Klang Valley is high as most of these organizations are committed as part of their CSR derives to reduce CO_2 and GHG. Being a high-income organization, the availability of internal financial funds to invest in programs to educate and train the employees on EE and EE implementation are promoted regularly (Theme #3, Theme #4).

Lack of EnMS Implementation. To successfully implement and sustain EE implementation, an organization should have an effective EnMS, but the research finding affirmed that the CIEC of Klang Valley who are of high-income, such as MNCs and large local companies, are the only ones who implement EnMS (Theme #6, Sub-theme #6a). In addition, a majority of them adopted the ISO 50001 standards (Theme #9, Sub-theme #9a). To implement and manage an EnMS ISO 50001 system requires high expenditure, and most of the CIECs of Klang Valley of low-income do not implement EnMS due to affordability (Marimon & Casadesús, 2017; Sub-theme #9a). The integration of EnMS into the overall management system of the CIECs of Klang valley is only carried on by high-income organizations. Furthermore, it is not mandatory to implement EnMS. The above-mentioned factor, due to the lack of EnMS affordability, is seen as one of the barriers restricting the adoption of EE technologies (Kelley, Goldberg, Magdon-Ismail, Mertsalov & Wallace, 2011; Kadam, 2014).

Lack of Government Support. This study found that support from Malaysian government entities responsible for energy efficiency and environmental protection only provide periodic and non-consistent assistance and collaboration in order to ensure that the CIECs of Klang Valley that come under the purview of these laws and regulation imposed by the Energy Commission of Malaysia are fulfilled (IEA, 2011). The only support currently provided to the CIECs of the Klang Valley are from private companies in the form of ESCOs who carry out energy performance contracting (EPC), which is a form of their business model for very basic EE implementation. For large-scale EE projects, ESCOs are unable to finance these project as it required a large upfront investment. According to Beuc (2014), large upfront costs (mainly for retrofits) still need to be integrated into a business model when demand for EE increases. These costs could be covered through energy savings by innovative financing in a form of EPC. This method offsets the EE investment cost against energy savings across the financing term, providing a zero-net-cost investment technique. ESCOs could leverage if the government encourages development banks to establish special purpose vehicles for buying a matured loan from them (Bache, 2014). Since EPC financing by banks are very rare, most high-income CIECs of Klang Valley do not really rely on government support as they have their own internal financial funding to implement EE measures. Those CIECs of Klang Valley with low-income are the ones who are affected by a lack of support from the government as they have insufficient internal funding to carry out EE measures. The lack of government support is one of the barriers as to why CIECs resist EE postures (Theme #1, Sub-theme #1a, Sub-theme #1b).

Affordability of Policies. The research concluded that the internal policies for reduction in CO_2 and GHG emission are normally carried out by high-income organizations, and in low-income organizations, it is absent. The purpose of high-income CIECs of Klang Valley in implementing internal policies has two objectives. First, to improve the company's image and brand. Second, to be socially responsible (Zaid, Myeda, Mahyuddin, &, Sulaiman, 2015; Theme #2, Sub-theme #2a, Sub-theme #2b). Implementing and managing these policies require large amounts of investment, and these companies are able to fund this internally due to their earning abilities. As for the low-income CIECs of Klang Valley, since they are constrained by the lack of internal funding, they are unable to implement the above-mentioned internal policies (Theme #3, Sub-theme #3a, Sub-theme #3b, Theme #4, Sub-theme #4a, Sub-theme #4b). Therefore, the affordability of policies is one of the barriers identified from this research as to why CIECs of Klang Valley resist adoption of EE technologies.

Lack of Information on EE Technologies. The published information on EE technologies currently available is only on low-hanging fruits EE measures. These basically comprise of light retrofitting such as replacing fluorescent lights to LEDs light system, replacement of conventional motors to high-efficiency motors, and the application of VSDs in process systems. Currently, these technologies are considered some of the best available technologies (BAT) providing rapid financial payback and high return on investment (Theme #5, Sub-theme #5a). There is a lack of published information or expertise for complex systems such as mechanical ventilation and air-conditioning (MVAC), vertical transportation, and electric induction boilers. These complex systems require customization in order to implement EE measures, have longer financial payback, and low return on investment (Theme #3, Sub-theme #3a, Sub-theme #3b). Most of the information provided by manufacturers for either BAT or for complex system EE do not cover information such as transactions and hidden costs that are not reflected in EE technologies engineering models. The lack of information on EE technologies is perceived as a barrier for why CIECs of Klang Valley resist the adoption of EE technologies (Backman, 2017; Carbon Trust, 2012; PWC, 2016; Srinivas et al., 2015).

Financial Scheme Availability. Most of the high-income CIECs of Klang Valley invest in EE implementation by utilizing internal financial funds, whereas low-income CIECs of Klang Valley normally seek external financial funding to implement EE technologies. These financial funds are in scarcity as the funds provided by government entities responsible for EE and environmental protection are limited (Theme #1, Sub-theme #1a). Financial institutes perceive EE projects as high-risk compared to conventional projects and are stringent in approving EE projects or impose higher interest rates that make some of the EE projects financially non-feasible (Srinivas et al., 2015). As pointed out by Apeaning (2012) in his research, the lack of understanding of EE technologies among the financial institute's personnel contributes to this issue. Therefore, the lack of financial schemes catering expressly for EE projects is not readily available (Apeaning, 2012).

RECOMMENDATIONS

The recommendation and fundamental obstacles for energy efficiency among the CIECs of Klang Valley are explained in detail below.

Though there are numerous energy and environmental laws and regulations implemented by the government of Malaysia through the agencies responsible to oversee the administration of sustainable energy and environmental protection, there seems to be a lack of supervision in enforcing these laws and regulations. Even though several building EE laws and policies show great objectives, their implementation faces many difficulties and resistances due to the behavioral attitude of CIECs toward EE adoption. As a result, these related agencies of the Malaysian government focus on supporting the implementation and supervision of the current energy and environmental protection laws and regulations. This study's findings concluded that there has been no evidence of the law being enforced such as fines or imprisonment. For the government to gauge the effectiveness of the laws and regulations implemented for the efficient management of electrical energy, it must carry out its own assessments by deploying research and survey teams to study the impact of the laws and regulations to create a system with continuous feedback to improve the laws, regulation, and enforcement mechanism. The Governance Assessment Tool (GAT) developed by Bressers et al. (2016) could be utilized by the Malaysian government to assess if the enforcement of regulation is carried out effectively. The GAT involves data collection by semi-structured interviews, reviews of policy documents, and secondary quantitative data. The first step is to establish the governance dimensions for the GAT tool. The following are the dimensions required to define relating with the efficient management of electrical energy laws and regulation of Malaysia:

- Levels and scales
- Actors and network
- Problem perspectives and goal ambitions
- Strategies and instruments
- Responsibilities and resources (Bressers, Bressers, Kuks & Larrue, 2016)

The second step is to establish the evaluation criteria as follows:

- Extent (are all relevant aspects taken into account?)
- Coherence (are all aspects reinforcing rather than contradicting each other?)
- Flexibility (are multiple roads to the goals, depending on opportunities and threats as they arise, allowed and supported?)
- Intensity (the degree to which the regime elements urge changes in the status quo or in current developments) (Bressers, Bressers, Kuks, & Larrue, 2016).

The third step is to combine the governance dimensions and evaluation criterion as a matrix (see Figure 24).

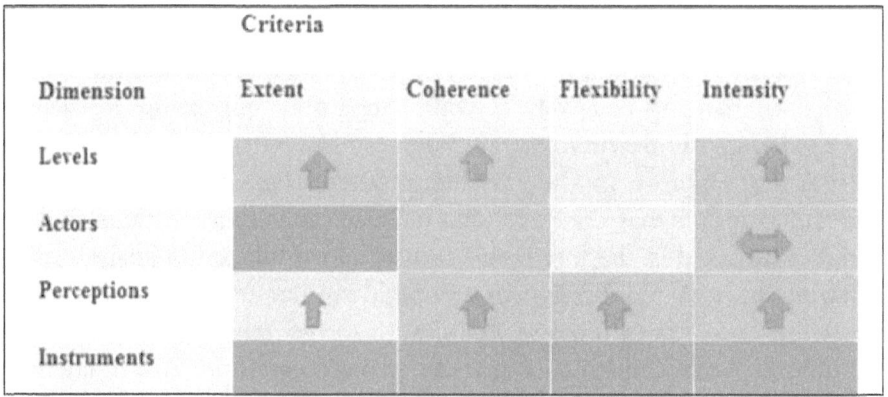

Figure 24 GAT scoring card visualization matrix. Adapted from "Adoption of Energy Efficiency Appliance by Households in Nigeria," by Ghana & Hope (2017).

The fourth step is to carry our data collection retrieved from different sources: interviews, text documents, and quantitative data. Face-to-face interviews should be conducted with sectorial experts such as the Ministry of Energy, Green Technology and Water (KeTTHA), Suruhanjaya Tenaga, Malaysia (Energy Commission, Malaysia), Malaysian Green Technology Corporation (MGTC), the Sustainability Energy Development Agency (SEDA), and selected energy experts. Semi-structured questionnaires consisting of open-ended questions could be utilized.

The fifth step is data analysis, where the GAT will be used to order, structure, and analyze the collected qualitative data. The analysis is to assess the governance of the efficient management of electrical energy laws and regulation of Malaysia. The interview transcripts will be analyzed by listing all the relevant answers by the sectorial experts interviewed. The frequencies of the different opinions per question will be translated into percentages, and the results of the analysis will be presented in a GAT scorecard (see Figure 24). Apart from the quantified type of analysis, qualitative interpretation of the data should be conducted as well.

The final step is the governance assessment guideline. A visual presentation of the governance of the energy efficiency policy is presented by means of a scorecard (See Figure 24). The green cells represent the results of the analyzed issues that are established as satisfactory according to the issue matrix (positive); red cells, on the contrary, present results that are worrying (negative); and orange cells present results that are rather unsatisfactory or uncertain (neutral). The up arrows indicate that the present situation is changing positively or will change positively in the foreseeable future, whereas the down arrows indicate that the situation is deteriorating and will not improve in the foreseeable future. Finally, the stable arrow (horizontal) means that the status quo or business as usual remains, and no change is expected in the foreseeable future. The results of an analysis using the above dimension and evaluation criteria will not tell one what the best policy option is; rather, it draws attention to the governance conditions that can hinder policy implementation and project execution under complex and dynamic conditions. The kind of policy advice the model generates is what barriers and hindrances a policy implementing such as Energy Commission of Malaysia need to cope with in the governance context and what conditions are, in this context, favorable to policy implementation and project execution (Ghana & Hoppe, 2017).

From the secondary data, the results from previous researches were analyzed using lifetime data. It is quite difficult to determine the future cost of electricity as the life-cycle cost approach in implementing EE technologies takes into account the energy saved discounted over the life of the technology employing Net Present Value (NPV) analysis, which poses a barrier in determining the actual success of the EE project. The sum of investment costs and the operating costs of the EE technology discounted over the lifetime (NPV) is taken to derive the return on investment (ROI) of the EE project. From the literature review and from the primary data of this research, it is established that EE projects aim at achieving the ROI within a three-year requirement, for example, a payback period and to fulfill an acceptable Internal Rate of Return (IRR) for the investment. The life cycle can comprise of a top-down or bottom-up approach, and such organizations can use either one of it based on the availability of data to select the best approach and be diligent on how the lifetime data analyzed is utilized. Realizable and realistic market potential cannot be termed the same. Realizable market potential is based on price, energy potential, and energy savings. Realistic market potential comprises non-financial barriers such as acceptance of solutions, and government entities such as the Energy Commission of Malaysia should provide proper calculation methodology or provide computational software where consumers of electricity could ascertain the realistic or achievable ROI before deciding on the implementation of the EE project.

According to the United Nations Economic Commission for Europe 2017, in order to motivate EE, there should be established standards, guidelines, and roadmaps suitable for the local environment. These have at present not been executed completely in Malaysia. As observed in this research, implementing EnMS and EE measures is not affordable by most CIECs, and government authorities must consider tax incentives, financial assistance, financial subsidies, and other schemes to promote energy efficiency development (Majumdar, 2010). The capabilities of the appropriate Malaysian government agencies in EE evaluation should be improved upon, and a market program offering EE solutions should be set up as the research findings clearly demonstrate that there is hardly any collaborative support or infrastructure provided by the government for EE to the CIECs of Klang Valley. Therefore, appropriate infrastructure to support EE should be set up by the Malaysian government, and one of its tasks is to study barriers that affect the proper financing of energy efficiency projects (Alcorta, Bazilian, De Simone & Pedersen, 2014).

There is a lack of comprehensive, authoritative, precise, and timely statistical information on building EE, which affects potential risk-taking in EE decision making or assessing from the appropriate segments of the government. Further, local EE statistics information and facts are difficult to attain. Hence, it is suggested that applicable divisions and organizations of the federal government should modify their attempts in gathering national and local EE statistical information and allow open access to these data as the interviews carried out in this research study concluded that the implementation of EE projects are seen as high-risk by financial institutes, thus not providing support to especially to ESCOs.

Presently, the EE management system and agencies in Malaysia are lagging behind. Promoting EE in the nation is difficult and also challenging since it generally requires the attention of numerous conflicting groups. It is suggested from the findings of this research that the strengths of the Ministry of Energy, Green Technology and Water (KeTTHA), Suruhanjaya Tenaga, Malaysia (Energy Commission, Malaysia), Malaysian Green Technology Corporation (MGTC), and Sustainability Energy Development Agency (SEDA) should all be merged so that this alliance could efficiently and effectively promote EE and amalgamate the leadership of Malaysia's energy efficiency.

Additionally, we could also encourage more research to be carried out to integrate the EE system, such as energy optimization systems for building services (into building design and operations). It is imperative and urgent that Malaysia implement energy efficiency building codes and further encourage stakeholders, such as the Board of Engineers, Malaysia (BEM), Institution of Engineers, Malaysia (IEM), and Pertubuhan Akitek Malaysia (PAM – Malaysian Institute of Architects), to mandate the involvement of energy consultants in all aspects of engineering, where the design involves energy utilization, especially in facilities and building design, in order to expedite energy saving effectiveness and to raise the industry's standards toward energy sustainability (Hubacek, Guan & Barua, 2007).

RECOMMENDATION FOR FUTURE RESEARCH

The implications for future research derived from the qualitative analysis of the present research study provide only the perceptions of 10 energy experts toward the adoption of EE technologies by CIEC of Klang Valley and the barriers associated with adoption of EE technologies. If the study follows a quantitative analysis, it would have identified other forms of barriers that resist the adoption of EE technologies, and it would provide a much more detailed analysis in regard to EE technologies adoption by CIEC of the Klang Valley. Furthermore, in a quantitative study, a large number of participants can be included to cover a larger group of energy experts in Klang Valley and outside of Klang Valley region, which would provide greater insights into EE technologies adoption, not only by CIEC of Klang Valley but Malaysia as a whole.

This study focused mainly on the commercial and industrial consumers of electricity toward the adoption of energy-efficient technologies since the largest consumers of electricity in most countries are from the commercial and industrial sectors. A separate future study should be carried out focusing on residential consumers of electricity and their attitudes and behaviors toward the adoption of energy-efficient technologies in the residential sector and how their attitudes and behaviors differ from CIECs. Since the government does not impose any form of regulation nor provides incentives to mandate residential consumers to implement energy-efficient technologies to reduce energy consumption in order to mitigate GHG and CO_2 emission, the findings may provide different insights. Furthermore, it is left entirely to the residential consumers to find their own finances and EE solutions to reduce electrical energy consumption and to overcome barriers in implementing energy-efficient technologies, as the engagement of energy experts will not be affordable by residential consumers of electricity.

This research study was carried out in the Klang Valley region; however, if the study covers other regions of Malaysia, new insights could be discovered on the various implementation techniques and policies developed in these other regions. This study could provide more knowledge in effective reduction of energy that could be further implemented in Malaysia. Additionally, data could be collected from other neighboring countries, such as ASEAN countries, to carry out a comparative analysis on laws and regulations, best available technologies (BATs), EnMS systems, implementation methods of EE, financial schemes for EE projects, methods to overcome barriers, and effective methods to enforce and maintain the laws and regulations pertaining to energy efficiency.

Finally, an essential extension of this research work will be to incorporate the views of external stakeholders like researchers, equipment dealers, financial organizations, local government, trade associations/unions, and many others on the barriers and drivers for improving commercial and industrial EE in the Klang Valley of Malaysia. By so doing, claims made by the research participants could be supported or refuted and an additional broad base of knowledge suitable for policy implementation will be developed. Lastly, studies could be narrowed down to high energy intensive and low energy intensive organizations.

References

ABB. (2017, July 14). *The state of global energy efficiency*. Retrieved from ABB Web site: http://www02.abb.com/global/uaabb/uaabb054.nsf/0/e06a0bb0a995c0e9c1257a410038260b/$file/The_state_of_global_energy_efficiency_Eng.pdf

Abdullah, H., Abu Bakar, N., Mohd Jali, M. R., & Ibrahim, F. W. (2017). The Current State of Malaysia's Journey towards a Green Economy: The Perceptions of the Companies on Environmental Efficiency and Sustainability. International Journal of Energy Ec. *International Journal of Energy Economics and Policy, 7*(1), 253-258.

ADB. (2011). *Malaysia: Supporting the Tenth Malaysian Plan, 2011–2015*. Tokyo: Asian Deveopment Bank. Retrieved June 2, 2017, from https://www.adb.org/projects/45083-001/main#project-pds

Adelman, C., Jenkins, D., & Kemmis, S. (1980). Rethinking Case Study: Notes from the Second Cambridge Conference. *Second Cambridge Conference* (pp. 45–61). Cambridge: University of East Anglia: Centre for Applied Research in Education.

Adger, W. N., Dessai, S., Goulden, M., Hulme, M., Lorenzoni, I., Nelson, D.,... Wreford, A. (2009). Are there social limits to adaptation to climate change? *Climatic Change, 93*(3–4), 335–354. doi:10.1007/s10584-008-9520-z

Akadiri, P., & Fadiya, O. (2013). Empricial analysis of the determinants of environmentally sustainable practices in UK construction industry. *Construction Innovation, 13*(4), 352–373. doi:10.1108/CI-05-2012-0025

Alcorta, L., Bazilian, M., De Simone, G., & Pedersen. (2014). Return on investment from industrial energy efficiency: evidence from developing countries. *Energy Efficiency, 7*(1), 43–53. doi:10.1007/s12053-013-9198

Alvesson, M., & Skoldberg, K. (2017). *Reflexive Methodology: New Vistas for Qualitative Research.* New Malden. Retrieved July 4, 2017, from https://books.google.co.in/books?id=9fI4DwAAQBAJ&dq=Rhetorical+assumption+qualitative+research&lr=&source=gbs_navlinks_s

Alvesson, M., & Skoldberg, K. (2017). Reflexive Methodology: New Vistas for Qualitative Research. New Malden. Retrieved May 7, 2018, from https://books.google.co.in/books?id=9fI4DwAAQBAJ&dq=Rhetorical+assumption+qualitative+research&lr=&source=gbs_navlinks_s

Anon. (2015). Philosophy and Phenomenolgical Research. *90(1)*, 1–256. Retrieved from http/onlinelibrary.wiley.com/doi/10.1111/phpr.2014.90.issue-1/issuetoc

Antwi, S., & Hamza, K. (2015). Qualitative and Quantitative Research Paradigms in Business Research: A Philosophical Reflection. *European Journal of Business and Management, 7*(3), 217–225.

APA. (2017). *APA Ethical Principle of Psychologist and Code of Conduct.* Retrieved September 15, 2017, from American Psychological Association Web site: www.apa.org/ethics/code/ethics-code-2017.pdf

Apeaning, R. (2012). *Energy Efficiency and Management in Industries – a case stusy of Ghana's largest industrial areas.* Master Thesis, Institute of Technology – Linkoping University – Division of Energy System. Retrieved July 15, 2017, from Diva-portal.org: https://www.diva-portal.org/smash/get/diva2:527775/FULLTEXT01.pdf

Apeaning, R., & Thollander, P. (2013). Barriers to and driving forces for industrial energy efficiency improvements in African industries: a case study of Ghana's largest industrial area. *Journal of Cleaner Production, 53*, 204-213. doi:10.1016/j.jclepro,2013.04.003

APEC. (2011). *Compendium of Energy Efficiency Policies of APEC Economies.* Tokyo: Asia Pacific Energy Research Centre, Institute of Energy Economics, Japan. Retrieved July 17, 2017, from www.ieej.or.jp/aperc

APEC. (2011). *Peer Review on Energy Efficiency in Malaysia.* Kuala Lumpur: Asia-Pacific Economic Cooperation. Retrieved June 15, 2017, from Asia-Pacific Economic Cooperation. (2011). Peer Review on Energy

Effhttps://www.ewg.apec.org/documents/9b_Malaysia PREE Report_Final Draft.pdf

APEC. (2011). *Peer Review on Energy Efficiency in Malaysia*. Asia-Pacific Economic Cooperation (APEC). Retrieved June 15, 2017, from https://www.ewg.apec.org/documents/9b_Malaysia PREE Report_Final Draft.pdf

Arasu, M., & Jeffrey, L. (2009). Energy Consumption Studies in Cast Iron Foundries. In Transcations of 57th IFC 2009. *57th Indian Foundry Congress* (pp. 330–336). Indian Foundry Congress. Retrieved August 31, 2017, from http://foundryinfo-india.org/images/pdf/57ifctp12.pdf

Argentina Country Reports. (2011). *Argentina Energy Efficiency Report*. Retrieved May 17, 2017, from Argentina Country Reports. (2011). Argentina Energrhttps://library.e.abb.com/public/47c05cea99dc331dc1257864004ccc8e/Argentina.pdf

Armel, C. (2014). *Energy Behavior*. Stanford University, Precourt Energy Efficiency Center. Washington: Advance Research Projects Agency – Energy – U.S. Department of Energy. Retrieved August 28, 2017, from https://arpa-e.energy.gov/sites/default/files/Armel Transport 03102014f.pdf

Auffhammer, M., Blumstein, C., & Fowlie, M. (2008). Demand-side Management and Energy Conservation Revisited. *The Energy Journal, 29*(3), 91–104. doi:10.2307/41823171

Ayres, R. U., & Warr, B. (2010). *The economic growth engine: How energy and work drive material prosperity* (Paperback ed.). Cheltenham, UK: Edward Elgar.

Azlina, A. A., Engku Abdullah, E., Kamaludin, M., & Radam, A. (2015). Energy Conservation Of Residential Sector In Malaysia. *Journal of Business and Social Development (Penjimatan Tenaga Bagi Sektor Kediaman Di Malaysia), 3*(2). Retrieved February 3, 2017, from http://jbsd.umt.edu.my/wp-content/uploads/sites/53/2015/09/4.-Pathway-Toward-web.pdf

Bache, A. (2014). Enabling Framework to Scale Up Investments in Energy Efficiency. *Commission on Environment and Energy*. Paris: International Chamber of Commerce. Retrieved August 7, 2017, from http://icc.nl/

docman-standpunten/docman-commissies/docman-commisies-klimaat-energie/58-icc-framework-to-scale-up-investments-in-energy-efficiency-1/file

Backman, F. (2017). Barriers to Energy Efficiency in Swedish Non-Energy-Intensive Micro – and Small-Sized Enterprises—A Case Study of a Local Energy Program. *Energies, 10*(1), 100. doi:doi: 10.3390/en10010100

Bank Negara Malaysia. (2015). *Statutory Requirements.* Retrieved June 15, 2017, from http://www.bnm.gov.my/files/publication/ar/en/2015/ar2015_book.pdf

Bazaley, P. (2009). Analysing qualitative data: More than 'Identifying Themes'. *The Malaysian Journal of Qualitative Research, 2*(2), 6–22.

BEE. (2017). *Energy Conservation Building Code (ECBC).* Retrieved October 11, 2017, from Bureau of Energy Efficiency, India: https://beeindia.gov.in/

Bolarinwa, O. (2015). Principles and methods of validity and reliability testing of questionnaries used in social and health science researches. *Nigeria Postgraduate Medical,* 195. doi:10.4103/1117-1936.173959.

Bressers, H., Bressers, N., Kuks, S., & Larrue, C. (2016). Governance for Drought Resilience. In The Governance Assessment Tool and Its Use. Germany: Springer.

Bryman, A. (2012). *Social Research Methods* (4th ed.). New York: Oxford University Press.

Bryman, B., & Bell, E. (2015). *Business Research Methods* (4th ed.). Oxford: Oxford University Press.

Cagno, E., & Trianni, A. (2013). Exploring drivers for energy efficiency within smalland medium-sized enterprises: First evidences from Italian manufacturing enterprises. *Applied Energy, 104,* 276–285.

Cagno, E., Worrell, E., Trianni, A., & Pugliese, G. (2013, March). A noval approach for barriers to industrial energy efficiency. *Renewable and Sustainable Energy Review, 19,* 230–308. doi:10.1016/j.rser.2012.11.007

Cambridge Econometrics. (2015). *Assessing the Employment and Social Impact of Energy Efficiency.* Warwick Institute. Cambridge: Warwick

Institute for Employment Research. Retrieved June 15, 2017, from http://ec.europa.eu/energy/sites/ener/files/documents/CE_EE_Jobs_main 18Nov2015.pdf

Camden. (2012). *Camden's behaviours framework Staff guide.* Camden: Camden Council. Retrieved July 1, 2017, from http://www.jobsgopublic.com/vacancy_attachments/215200?source=att · PDF file

Capgemini. (2011). *Capgemini.* Retrieved August 28, 2017, from Capgemini website: https://www.capgemini.com/resource-file-access/resource/pdf/Digital_Transformation__A_Road-Map_for_Billion-Dollar_Organizations.pdf

Carbon Trust. (2012, March). *Metering: Introducing the techniques and technology for energy data management.* Retrieved August 26, 2017, from The Carbon Trust Website: https://www.carbontrust.com/media/31679/ctv027_metering_technology_overview.pdf

Carbon Trust. (2018, April). *Employee Awareness and Office Energy Efficiency.* Retrieved May 1, 2018, from Carbob Trust Web site: https://www.carbontrust.com/resources/guides/energy-efficiency/employee-awareness-and-office-energy-efficiency/#docdownload-1479

Carley, S., & Lawrence, S. (2014). Carley, S., & Lawrence, S. (2014). Defining Energy-Based Economic Development. In Energy-Based Economic Development (pp. 15–35). London: Springer London. doi:10.1007/978-1-4471-6341-1_2.

CCES. (2010). *Barriers and Solutions to Corporate Energy Efficiency.* Retrieved July 13, 2017, from Centre for Climate and Energy Solutions Web Site (CCES): https://www.c2es.org/docUploads/Barriers and Solutions to Energy Efficiency Chart.pdf

Chai, K.-H., & Yeo, C. (2012). Overcoming energy efficiency barriers through systems approach—A conceptual framework. *Energy Policy, 46,* 460–472. doi:10.1016/j.enpol.2012.04.012

Chan, E. H., Qian, Q. K., & Lam, P. T. (2009). The market for green building in developed Asian cities—the perspectives of building designers. *Energy Policy, 37,* 3061–3070. doi:10.1016/j.enpol.2009.03.057

Chpassociation. (2015). *Barrier to Industrial Energy Efficiency: A Study Pursuant to Section 7 of the American Energy Manufacturing Technical*

Corrections Act. Washington: U.S. Department of Energy. Retrieved July 11, 2017, from http://chpassociation.org/wp-content/uploads/2015/06/Barriers-to-Industrial-Energy-Efficiency-Study_June-20

Cohen, L., Manion, L., & Morrison, K. (2000). *Research Methods in Education* (5th ed.). London and New York: Routledge.

Collis, J., & Hussey, R. (2013). *A Practical Guide for Undergraduate and Postgraduate Students* (4th ed.). United Kingdom: Palgrave Macmillan.

Commission of the European Communities. (2005). *Green Paper on Energy Efficiency or Doing More With Less, COM (2005) 265 final*. Brussels: Commission of the European Communities. Retrieved August 16, 2017, from www02.abb.com/.../European+Union+Green+Paper+on+Energy+Efficiency.pdf · PDF file

Copenhagen Centre on Energy Efficiency. (2015). *Accelerating Energy Efficiency: Initiatives and Opportunities – Southeast Asia*. Retrieved July 18, 2017, from Copenhagen Centre on Energy Efficiency Web Site: http://kms.energyefficiencycentre.org/sites/default/files/C2E2_

Crafts, N. F. (2004). *The World Economy In The 1990s: A Long Run Perspective*. Working Paper No. 87/04, London School of Economics, Department of Economic History, London. Retrieved July 20, 2017, from http://eprints.lse.ac.uk/22334/1/WP87.pdf

Creswell, J. (2014). *Research Design: Qualitative, Quantitative, & Mixed methods approaches* (4th ed.). London: SAGE Publications Inc.

Crossman, A. (2017). *The Case Study Research Method Definition and Different Types*. Retrieved from https://www.thoughtco.com/case-study-definition-3026125

Crotty, M. (2003). *The Foundations of Social Research: Meaning and Perspectives in the Research Process* (3rd ed.). London: Sage Publications.

Davies, P., & Osmani, M. (2011). Low carbon housing refurbishment challenges and incentives: Architects' perspectives. *Built Environment, 46*, 1691–1698.

Denzin, N., & Lincoln, Y. (2017). *Handbook of qualitative rearch* (5th ed.). Thousand Oaks, CA: SAGE.

Department of Statistics Malaysia. (2017). *Federal Territory of Kuala Lumpur*. Retrieved June 20, 2017, from Department of Statistics Malaysia Web

Site: https://www.dosm.gov.my/v1/index.php?r=column/cone&menu_id=bjRlZXVGdnBueDJKY1BPWEFPRlhIdz09

DOE. (2015a). *Barriers to Industrial Energy Efficiency.* Washington D.C: U.S. Department of Energy. Retrieved August 2, 2017, from https://www.gpi.org/sites/default/files/Barriers to Industrial... ·PDF file

DOE. (2015b). *Quadrennial Technology Review 2015: Increasing Efficiency of Building Systems and Technologies Supplemental Information.* U.S. Department of Energy (DOE), Washington. Retrieved June 17, 2017, from https://energy.gov/sites/prod/files/2015/10/f27/Ch5

Dudovskiy, J. (2015). *The Ultimate Guide to Writing A Dissertation in Business Studies – A Step-by-Step Assistance.* London, U.K: research-methodology.net.

Dunstan, C., Daly, J., Langham, E., Boronyak, L., & Rutovitz, J. (2011). *Intelligent Grid Research Cluster – Project 4 Institutional Barriers, Economic Modelling and Stakeholder Engagement: Institutional Barriers to Intelligent Grid (No. 4.1).* Institute for Sustainable Futures. Igrid. Retrieved August 2, 2017, from http://igrid.net.au/resources/downloads/project4/P4 I

Eang, L. S. (2015). *A Review of Building Energy Efficiency Development in Malaysia. Retrieved from.* Retrieved August 9, 2017, from wbcsd resources Web Site: http://wbcsdpublications.org/wp-content/uploads/2015/12/MarketReview-Malaysia.pdf

Easterby-Smith, M., Lewis, P., & Thornhill, A. (2012). *Research Methods for Business Student* (6th ed.). Harlow: Pearson Education Limited.

EC-Europa. (2017). *Study evaluating progress in the implementation of Article 7 of the Energy Efficiency Directive.* DG Energy. Oxford: Ricardo Energy & Environment. Retrieved August 30, 2017, from https://ec.europa.eu/energy/sites/ener/files/documents/final_report_evaluation_on_implementation_art._7_eed

EEA. (2017). *EN17 Total Energy Intensity.* European Environment Agency. Retrieved from https://www.eea.europa.eu/data-and-maps/indicators/en17-total-energy-intensity/en17-total-energy-intensity

EIA. (2014). *Annual Energy Outlook 2014 with Projections to 2040.* Energy Information Administration (EIA). Retrieved August 30, 2017, from https://www.eia.gov/outlooks/aeo/pdf/0383(2014).pdf

Ekstrom, J. A., Moser, S. C., & & Torn, M. (2011). *Barriers to Climate Change Adaption: A Diagnostic Framework.* Berkeley: California Energy Commission. Retrieved July 21, 2017, from http://www.energy.ca.gov/2011publications/CEC-500-2011-004/CEC-500-2011-004.pdf

Energy Commission of Malaysia. (2008). *An Overview of the Efficient Management of Electrical Energy Regulation 2008.* Retrieved May 7, 2018, from st.gov.my Web site: https://www.st.gov.my/index.php/en/applications/energy-efficiency/registered-electrical-energy-managers?id=141

Energy Commission of Malaysia. (2014). *National Energy Balance 2013.* Putrajaya: Suruhanjaya Tenaga, Malaysia (Energy Commission of Malaysia). Retrieved July 2, 2017, from http://meih.st.gov.my/documents/10620/167a0433-510c-4a4e-81cd-fb178dcb156f· PDF file

Energy Commission of Malaysia. (2016). *Malaysia Energy Information Hub-statistics.* Putrajaya: Suruhanjaya Tenaga, Malaysia (Energy Commission of Malaysia). Retrieved July 2, 2017, from http://meih.st.gov.my/statistics

Energy Commission of Malaysia. (2018, May 1). *List of Registered Electrical Energy Managers.* Retrieved from Energy Commission of Malaysia Web site: http://ecos.st.gov.my/ms/web/guest/senarai-pengurus-tenaga-elecktrik-berdaftar

Energy Community. (2017). *Energy efficient building implementation.* Retrieved July 11, 2017, from Energy Community Web Site: http://www.energycommunity.org/documents/SEA6.Energy efficient building implementation.pdf

Enfield Council. (2017). Leadership Competencies Framework. Enfield, UK. Retrieved June 15, 2017, from www.enfield.gov.uk/download/downloads/id/9485/leadership_competencies·PDF file

Environmental Defence Fund. (2017). EDF Climate Corps Handbook. *Strategic Energy Management for Organization*(8). New York, New York, USA. Retrieved July 15, 2017, from http://edfclimatecorps.org/sites/edfclimatecorps.org/files/edfclimatecorps_handbook_8th ed.pdf

EPA. (2009). *Energy Efficiency as a Low-Cost Resource for Achieving Carbon Emission Reductions.* U.S. Environmental Protection Agency.

Retrieved June 15, 2017, from https://www.epa.gov/sites/production/files/2015-08/documents/ee_a

EPA. (2017). *Energy Efficiency in Local Government Operations*. Washington D.C.: U.S. Environmental Protection Agency (EPA). Retrieved July 14, 2017, from https://www.epa.gov/sites/production/files/2015-08/documents/ee_municipal_operations.pdf

ERE Consulting Group. (2017). *Mass Rapid Transport Project: Detailed Environmental Impact Assessment*. Retrieved October 11, 2017, from ere consulting group Web Site: http://ere.com.my/projects/klang-valley-mrt-sg-buloh-kajang-line/

European Commision. (2016). *Overview of support activities and projects of the European Union on energy efficiency and renewable energy in the heating and cooling sector*. Luxembourg: European Commission. Retrieved August 28, 2017, from http://ec.europa.eu/energy/sites/ener/files/document

European Commision. (2017). *Overview of support activities and projects of the European Union on energy efficiency and renewable energy in the heating and cooling sector*. European Commission. Retrieved August 28, 2017, from http://ec.europa.eu/energy/sites/ener/files/document

European Commission. (2007). *Energy Efficient Technology: Educational material TreeSpa Workshop*. Retrieved August 25, 2017, from European Commission Web Site: https://ec.europa.eu/energy/intelligent/projects/sites/iee-projects/files/projects/documents/treespa_energy_efficient_t

Fawkes, S., Oung, K., & Thorpe, D. (2016). *Best Practices and Case Studies for Industrial Energy Efficiency Improvement – An Introduction for Policy Makers*. Copenhagen: UNEP DTU Partnership.

Fleiter, T., Schleich, J., & Ravivanpong, P. (2012). Adoption of EnergyEfficiency Measures in SMEs–An Empirical Analysis Based on Energy Audit Data from Germany. *Energy Policy, 51*, 863–875. Retrieved September 20, 2017, from http://publica.fraunhofer.de/dokumente/N-219430.htm

Foxon, T. J. (2011). A coevolutionary framework for analysing a transition to a sustainable low carbon economy. *Ecological Economics, 70*(12), 2258–2267. doi:10.1016/j.ecolecon.2011.07.014

Foxon, T. J., & Steinberger, J. (2013). *Energy, efficiency and economic growth: a coevolutionary perspective and implications for a low carbon transition.* Sustainability Research Institute, Centre for Climate Change Economics and Policy – University of Leeds & London School of Economics. Leeds: Sustainabiltiy Research Insititute. Retrieved from http://sure-infrastructure.leeds.ac.uk/enecon

Gareis, R., Heumann, M., & Martinuzzi, A. (2009). *Relating sustainable develpoment and project management.* Brelin: IRNOP IX.

GSEP. (2013). *Knowledge and Skills Need to Implement Energy Management Systems in Industry and Commercial Buildings.* Clean Energy Ministerial (CEM), Global Superior Energy Performance Partnership. Global Superior Energy Performance Partnership. Retrieved June 10, 2017, from http://www.cleanenergyministerial.org/Portals/2/pdfs/GSEP_knowledge_skills_EnMS_implementation.pdf

Haig, B. D. (2018). *The Philosophy of Quantitative Methods: Understanding Statistics.* Oxford: Oxford University Press.

Hannes, K. (2011). Chapter 4: Critical appraisal of qualitative research. In J. Noyes, A. Booth, K. Hannes, A. Harden, J. Harris, S. Lewin, & C. Lockwood, *Supplementary guidance for inclusion of qualitative research in Cochrane systematic reviews of interventions (Version 1).* Cochrane Collaboration Qualitative Methods Group.

Harvey, L. (2015). Beyond member-checking: a dialogic apprach to the research interview. *International Journal of Research & Method in Education, 38*(1), 23–38.

Hassani, H. (2017). *Research Methods in Computer Science: The challenges and Issues.* Retrieved November 22, 2017, from arxiv.org: http://arxiv.org/pdf/1703.04080.pdf

Hesse-Biber, S. L. (2011). *The Practice of Qualitative Research* (2nd ed.). London, UK: SAGE Publications.

Hezri, A. A. (2016). *The Sustainability Shift: Refeshioning Malaysia's Future.* Penang, Malaysia: Areca Books.

Hiller, J., Mills, V., & Reyna, E. (2012). *Breaking Down Barriers to Energy Efficiency.* Retrieved June 23, 2017, from Ingersollrand Web Site: http://www.cees.ingersollrand.com/CES_documents/EDF_Breaking_Down_EE_Barriers.April2012.pdf

Huang, E. G. (2011). *Understanding the Requirements of the Energy Management System Certification.* SGS. SGS. Retrieved July 14, 2017, from http://www.sgs.com/-/media/Global/Documents/White Papers/sgs-energy-management-whitepaper-en-11.ashx

Hubacek, K., Guan, D., & Barua, A. (2007). Changing lifestyles and consumption patterns in developing countries: A scenario analysis for China and India. *Futures, 39*(9), 1084–1096. Retrieved July 14, 2017, from https://webcache.googleusercontent.com/search?q=cache:f4hTC

ICC India. (2014). *Enabling Framework to Scale Up Investments in Energy Efficiency: Commission on Environment and Energy.* Retrieved August 30, 2017, from International Chamber of Commerce India Web Site: http://www.iccindiaonline.org/policy-statement/may2014/5.pdf

ICF Consulting Limited. (2012). *Study on Energy Efficiency and Energy Saving Potential in Industry and On Possible Policy Mechanisms.* ICF Consulting Limited. London: ICF Consulting Limited.

IEA. (2010). *Technology Roadmap Energy-efficient Buildings: Heating and Cooling Equipment.* International Energy Agency. Retrieved August 28, 2017, from https://www.iea.org/publications/freepublications/publication/buildings_roadmap.pdf

IEA. (2011). *25 Energy Efficiency Policy.* International Energy Agency. EIA. Retrieved July 11, 2017, from https://www.iea.org/publications/freepublications/publication/25recom_2011.pdf

IEA. (2013). *IEA Annual Report.* International Energy Agency (EIA). EIA. Retrieved February 16, 2017, from https://www.iea.org/publications/.../2013_AnnualReport.pdf

IEA. (2016a). *Energy Efficiency Market Report.* International Energy Agency. International Energy Agency. Retrieved June 23, 2017, from https://www.iea.org/eemr16/files/medium-term-energy-efficiency-2016_WEB.PDF

IEA. (2016b). *World Energy Outlook 2016.* International Energy Agency. International Energy Agency. Retrieved August 30, 2017, from https://www.iea.org/publications/freepublications/publication/WorldEnergyOutlook2016ExecutiveSummaryEnglish.pdf

IISD. (2013). *A Citizens' Guide to Energy Subsidies in Malaysia.* The International Institute for Sustainable Development (IISD). Geneva:

IISD. Retrieved February 5, 2018, from http://iisd.org/gsi/sites/default/files/ffs_malaysia_czguide.pdf

Index Mundi. (2016). *Malaysia Demographics Profile.* Retrieved June 20, 2017, from Index Mundi Web Site: http://www.indexmundi.com/malaysia/demographics_profile.html

Inside Investor. (2012). *Greater Kuala Lumpur & Klang Valley.* Retrieved June 20, 2017, from etp.pemandu.gov.my Web Site: http://etp.pemandu.gov.my/upload/Inside Investor – Greater KL and Klang Valley.pdf

inta-aivn. (2017). INTA International Urban Development Association. Paris, France. Retrieved January 20, 2017, from http:/www.inta-aivn.org

IPEEC. (2016). *G20 Energy Efficiency Leading Programme.* Retrieved July 14, 2017, from International Partnership for Energy Efficiency Cooperation (IPEEC) Web Site: https://ipeec.org/upload/publication_related_language/pdf/481.pdf

Irrek, W., & Thomas, S. (2008). *Defining Energy Efficiency.* Retrieved April 24, 2017, from Wuppertal Institut Web Site: https://wupperinst.org/uploads/tx_wupperinst/energy_efficiency_definition.pdf

ISO. (2011). *Energy management.* Retrieved January 9, 2018, from International Organization for Standardization Web Site: https://www.iso.org/iso-50001-energy-management.html.

Kadam, S. V. (2014). Barriers in implementation of Energy Efficient Technologies in Selected Automobile Industries in Pune. *International Journal of Scientific Reserach and Management (IJSRM), 2*(9), 1409–1416.

Kelley, S., Goldberg, M., Magdon-Ismail, M., Mertsalov, K., & Wallace, A. (2011). Defining and discovering communities in social networks. In S. Kelley, M. Goldberg, M. Magdon-Ismail, K. Mertsalov, & A. Wallace, *Handbook of Optimization in Complex Networks* (pp. 139–168). Berlin: Springer.

KeTTHA. (2014). *National Energy Efficiency Action Plan: Draft Final Report.* Kementerian Tenaga, Teknologi Hijau dan Air (Kettha) (Ministry of Energy, Green Technology and Water). Putrajaya: Kettha. Retrieved

October 11, 2017, from http://www.kettha.gov.my/kettha/portal/document/files/NEEAP For Comments Final January 2014.pdf

KeTTHA. (2017). *EE Challenge 2014 Award Ceremony and Seminar on EPC Implementation in Government Buildings: Government Initiatives on Energy Efficiency in Malaysia.* Retrieved from Ministry of Energy, Green Technology and Water Web Site: www.st.gov.my/index.php/en/download-page/category/120-seminar-on... · PDF file

Khairunnisa, A. R., Yusof, M. Z., Salleh, M. N., & Leman, A. M. (2015). The Development of Energy Efficiency Estimation System (EEES) for Sustainable Development: A Proposed Study. *Energy Procedia.* doi:10.1016/j.egypro.2015.11.527

Khosravani, H., Castilla, M., M., B., Ruano, A., & Ferreira, P. (2016). A Comparison of Energy Consumption Prediction Models Based on Neural Networks of a Bioclimatic Building. *Energies, 9*(1), 57. doi:10.3390/en9010057

Kill, J., Ozinga, S., Pavett, S., & Wainwright, R. (2010, August). *Trading Carbon: How is works and why it is controversial.* Retrieved December 31, 2017, from FERN: http: www.fern.org/sites/fern,org/files/tradingcarbon_internet_Final.pdf

Kingdon, C. (2014). *Sociology for Midwives.* London, UK: Andrews UK Limited.

Koen, V., Asada, H., Nixon, S., Habeeb Rahman, M. Z., & Mohd Arif, A. Z. (2017). *Malaysia's Economic Success Story And Challenges (Economics Department Working Papers No. 1369).* Organisation for Economic Co-operation and Development (OECD), Economic Department. OECD. Retrieved June 10, 2017, from https://www.oecd.org/eco/Malaysia-s-economic-success-story-and-challenges.pdf

Kostka, G., Moslener, U., & Andreas, J. (2013). Barriers to increasing energy efficiency: evidence from small-and medium-sized enterprises in China. *Journal of Cleaner Production, 57,* 59–68.

Kusek, J. Z., & Rist, R. C. (2004). *Ten Steps to a Results Based Monitoring and Evaluation System.* Washington, D.C., USA: The World Bank.

Labuschagne, C., & Brent, A. (2006). Social indicators for sustainable project and techmology life cycle management in the process industry. *International Journal of Life Cycle Assessment, 11*(1), 3–15.

Lapan, S., Quartaroli, M., & Riemer, F. (2012). *Qualitative Research: An Introduction to Methods and Designs* (1st ed.). San Francisco, USA: Jossey-Bass Publishing.

Leung, L. (2015). Validity, realiabilty, and generalizability in qualiative research. *Journal of Family Medicine and Primary Care, 4*(3), 324. Retrieved October 22, 2017, from http://www.jfmpc.com/text.asp?2015/4/3/324/161306.

Levy, D. (2006). Qualitative Methodology and Grounded Theory in Property Research. *Pacific Rim Property Research Journal, 12*(4), 369–388. doi:10.1080/14445921.2006.11104216

Li, G., Xu, Z., Xiong, C., Yang, C., Zhang, S., Chen, Y., & Xu, S. (2011). Energy-efficient wireless communications: tutorial, survey, and open issues. *IEEE Wireless Communications, 18*(6), 28–35. doi:10.1109/MWC.2011.6108331

Lincoln, Y., & Guba, E. (1985). *Naturalistic Inquiry.* Beverly Hills, Calif.: SAGE Publications.

Ling, L., & Ling, P. (2015). *Methods and Paradigms in Education Research.* USA: IGI Global.

Linguistics Association. (2017). *Saudi Students' Social Identity and their Identity in Academic Writing: A Qualitative Study of Saudi Students in the Uk.* Leicester: University of Leicester. Retrieved July 8, 2017, from https://lra.le.ac.uk/bitstream/2381/39950/1/2017TajSRPhD.pdf

Lobe, B., Livingstone, S. O., & Simões, A. (2008, October 25). *Best Practice Research Guide: How to research children and online technologies in comparative perspective.* London: EU Kids Online.

Loftus, S., & Higgs, J. (2010). Researching the individual in workplace research. *Journal of Education and Work, 23*(4), 377–388. doi:10.1080/13639080.2010.495712.

Lopez, G. (2016). *What Kind of High-Income Country will Malaysia Become? Brinknews. Retrieved from.* Retrieved May 4, 2018, from Brinknews Web site: http://www.brinknews.com/asia/what-kind-of-high-income-country-will-malaysia-become/

Lunt, P., Ball, P., & Levers, A. (2014). Barriers to industrial energy efficiency. *International Journal of Energy Sector Management, 8*(3), 380–394. doi:10.1108/IJESM-05-2013-0008

Majumdar, M. (2010). *Green Buildings and their Financial Feasibility.* The Energy and Resource Institute (TERI), India. Retrieved August 28, 2017, from http://naredco.in/pdfs/Mili-Majumdar.pdf.

Malay Mail Online. (2017). *ETP 2013 Annual Report — Pemandu.* Retrieved June 20, 2017, from Malay Mail Online Web Site: Retrieved June 20, 2017, from http://www.themalaymailonline.com/what-you-think/article/etp-2013-annual-report-pemandu.

Marimon, F., & Casadesús, M. (2017). Reasons to Adopt ISO 50001 Energy Management System. *Sustainability, 9*(10), 1740-1755. doi:10.3390/su9101740

Marshall, C., & Rossman, G. B. (2014). *Designing Qualitative Research* (6th ed.). Raisan: SAGE Publications.

Matthews, B., & Ross, L. (2010). *Research methods: a practice guide for the social sciences.* Harlow: Longman. Retrieved October 22, 2017, from http://prism.tails.com/derby-ac/

May, T. (2011). *Social research: issues, methods and process.* Maidenhead: Open University Press. Retrieved October 22, 2017, from http://prism.talis.com/derby-ac/

McKane, A., Scheihing, P., & Williams, R. (2007). *Certifying Industrial Energy Efficiency Performance: Aligning Management, Measurement, and Practice to Create Market Value.* Berkeley: American Council for an Energy Efficient Economy (ACEEE). Retrieved November 30, 2017, from http://aceee.org/files/proceedings/2007/data/papers/56_5

McKinsey on Society. (2010). *Energy Efficiency: A compelling global resourse.* Retrieved July 14, 2017, from McKinsey on Society Web Site: http://mckinseyonsociety.com/energy-efficiency-a-compelling-global-resource/.

Meih. (2016). *Malaysia Energy Statistic Handbook.* Retrieved June 20, 2017, from Malaysia Energy Information Hub Unit (Meih): http://meih.st.gov.my/documents/10620/57af5e2a-7695-4618-a111-4ba0a49ba992.

Meih. (2017). *Introduction to Malaysia Energy Information Hub.* Retrieved October 20, 2017, from Meih Web Site: http://meih.st.gov.my/

Menezes, A. C., Cripps, A., Buswell, R. A., Wright, J., & Bouchlaghem, D. (2014). Estimating the energy consumption and power demand of small power equipment in office buildings. *Energy and Buildings, 75*, 199–209. doi:10.1016/j.enbuild.2014.02.011

Merriam, S., & Tisdell, E. (2016). *Qualitative Research: A Guide to Design and Implementation* (4th ed.). San Francisco, CA: Jossey-Bass.

Michaelides, R., Bryde, D., & Ohaeri, U. (2014). Sustainability from a project management perspective: are oil and gas supply chains ready to embed sustainability in their projects? *Project Management Institute Research and Education Conference, Phoenix, AZ.* Newtown Square, PA: Project Management Institute.

MIDA. (2017). *Malaysian Investment Development Authority.* Retrieved January 9, 2017, from MIDA Web site: http://www.mida.gov.my/home/

Montalvo, C. (2008). General wisdom concerning the factors affecting the adoption of cleaner technologies: a survey 1990–2007. *Journal of Cleaner Production, 16*(1), S7-S13.

Murphy, D., & Harris, M. (2014). *Policy and Regulatory Barriers in Kenya's National Climate Change Action Plan: Areas for Private Sector Action.* Retrieved July 13, 2017, from https://cdkn.org/wp-content/uploads/2015/04/Policy-Regulation-Review.pdf.

Myers, M., & Newman, M. (2007). The qualitative interview in IS research: Examining the craft. *Information & Orgnization, 17*(1), 2–26.

New Enfield. (2017). *Staff Competencies Framework.* Retrieved August 31, 2017, from New Enfield Web Site: https://new.enfield.gov.uk/services/jobs-and-careers/working-with-us/our-ways-of-working/working-with-us-information-staff-competencies.pdf

Noka Group. (2017). *Energy Efficiency for Sustainable Development.* Retrieved July 14, 2017, from Noka Group Web site: http://nokagroup.com/wp-content/uploads/2016/10/brochure_noka_inglese-ilovepdf-compressed.pdf

NPIC. (2017). *National Property Information Centre: Key Statistics.* Retrieved October 11, 2017, from National Property Information Centre (NPIC), Malaysia Web site: http://napic.jpph.gov.my/portal/main-page?p_p_id=ViewPublishings_WAR_ViewPublishingsportlet&p_p_lifecycle=0&p_p_state=norm

O'Brien, D., & Scot, P. (2012). Correlation and Regression. In H. Chen, *Approaches to Quantitative Research – A Guide for Dissertation Students* (pp. 122–133). Cork, Ireland: Oak Tree Press.

Observatorio Asia Pacifico. (2016). *Eleventh Malaysia Plan.* Observatorio Asia Pacifico. Observatorio Asia Pacifico. Retrieved June 20, 2017, from http://www.observatorioasiapacifico.org/data/OBSERVATORIO.Images/Bulletin/60/20160629020712EleventhMalaysiaPlan_ENGLISH.pdf.

OECD. (1996). *The Knowledge-Based Economy.* Organization for Economic Co-Operation and Development (OECD). OECD. Retrieved August 26, 2017, from https://www.oecd.org/sti/sci-tech/1913021.pdf.

OECD. (2011). *A Skilled Workforce for Strong, Sustainable and Balanced Growth: A G20 Training Strategy.* Organization for Economic Co-Operation and Development (OECD). OECD. Retrieved August 26, 2017, from https://www.oecd.org/g20/summits/toronto/G20-Skills-Strategy.pdf.

OECD. (2015). *An Introduction to Energy Management Systems: Energy Savings and Increased Industrial Productivity for the Iron and Steel Sector.* Organisation for Economic Co-operation and Development (OECD). OECD. Retrieved July 13, 2017, from OECD Web site: https://www.oecd.org/sti/ind/DS

OECD. (2015). *Energy Efficiency in the Steel Sector: Why It Works Well, But Not Always.* Organization for Economic Co-Operation and Development (OECD). OECD. Retrieved August 28, 2017, from https://www.oecd.org/sti/ind/Energy-efficiency-steel-sector-1.pdf.

OECD. (2017). *Competency Framework.* Organization for Economic Co-Operation and Development (OECD). OECD. Retrieved August 1, 2017, from https://www.oecd.org/careers/competency_framework_en.pdf.

OEE. (2012). *Implementing an Energy Efficiency Awareness Program.* Office of Energy Efficiency (OEE). OEE. Retrieved July 11, 2017, from https://oee.nrcan.gc.ca/sites/oee.nrcan.gc.ca/files/files/pdf/publications/commercial/Awareness_Program_e.pdf.

Olivier, J. G., Janssens-Maenhout, G., Muntean, M., & Peters, J. A. (2015). *Trends in global CO2 emissions: 2015 Report.* PBL Netherlands

Environmental Assessment Agency. The Hague: European Commission, Joint Research Centre; Ispra.

Oppong, S. (2013). The Problem of Sampling in Qualitative Research. *Asian Journal of Management Sciences and Education, 2*(2), 202–210.

Osmani, M., & O'Reilly, A. (2009). Feasibility of zero carbon homes in England by 2016: a house builder's perspective. *Build Environment, 44*, 1917–1924.

Othman, J., & Yahoo, M. (2014). Reducing CO2 Emission in Malaysia: Do Carbon Taxes Work? *9th PERKEM Proceedings* (p. 175). National University of Malaysia. Retrieved February 4, 2018, from http://www.ukm.my/fep/perkem/pdf/perkem2014/PERKEM_2014_ID4.pdf

P1. (2017, October 27). Perception of Energy Experts on Commerical and Industrial Electricity Consumers (CIEC) of the Klang Valley towards the Adoption of Energy Efficienct Technologies. (T. Subramaniam, Interviewer) Puchong Jaya, Selangor D.E., Malaysia.

P10. (2017, November 25). Perception of Energy Experts on Commerical and Industrial Electricity Consumers (CIEC) of the Klang Valley towards the Adoption of Energy Efficienct Technologies. (T. Subramaniam, Interviewer) Cyberjaya, Selangor D.E., Malaysia.

P2. (2017, October 25). Perception of Energy Experts on Commerical and Industrial Electricity Consumers (CIEC) of the Klang Valley towards the Adoption of Energy Efficienct Technologies. (T. Subramaniam, Interviewer) Puchong Jaya, Selangor D.E., Malaysia.

P3. (2017, October 29). Perception of Energy Experts on Commerical and Industrial Electricity Consumers (CIEC) of the Klang Valley towards the Adoption of Energy Efficienct Technologies. (T. Subramaniam, Interviewer) Shah Alam, Selangor D.E., Malaysia.

P4. (2017, November 5). Perception of Energy Experts on Commerical and Industrial Electricity Consumers (CIEC) of the Klang Valley towards the Adoption of Energy Efficienct Technologies. (T. Subramaniam, Interviewer) Dengkil, Selangor D.E., Malaysia.

P5. (2017, November 5). Perception of Energy Experts on Commerical and Industrial Electricity Consumers (CIEC) of the Klang Valley towards the Adoption of Energy Efficienct Technologies. (T. Subramaniam, Interviewer) Petaling Jaya, Selangor D.E., Malaysia.

P6. (2017, November 16). *Perception of Energy Experts on Commerical and Industrial Electricity Consumers (CIEC) of the Klang Valley towards the Adoption of Energy Efficienct Technologies*. (T. Subramaniam, Interviewer) Bandar Baru Bangi, Selangor D.E., Malaysia.

P7. (2017, November 14). *Perception of Energy Experts on Commerical and Industrial Electricity Consumers (CIEC) of the Klang Valley towards the Adoption of Energy Efficienct Technologies*. (T. Subramaniam, Interviewer) Shah Alam, Selangor D.E., Malaysia.

P8. (2017, November 21). *Perception of Energy Experts on Commerical and Industrial Electricity Consumers (CIEC) of the Klang Valley towards the Adoption of Energy Efficienct Technologies*. (T. Subramaniam, Interviewer) Kelana Jaya, Sealngor D.E, Malaysia.

P9. (2017, November 24). *Perception of Energy Experts on Commerical and Industrial Electricity Consumers (CIEC) of the Klang Valley towards the Adoption of Energy Efficienct Technologies*. (T. Subramaniam, Interviewer) Petaling Jaya, Selangor D.E., Malaysia.

Park, N.-K., & Lee, E. (2013). Energy-Efficient Lighting: Consumers' Perceptions and Behaviors. *International Journal of Marketing Studies, 5*(3). doi:10.5539/ijms.v5n3p26.

PEMANDU. (2012). *Greater Kuala Lumpur AND Klang Valley*. Retrieved June 15, 2017, from PEMANDU Web site: http://etp.pemandu.gov.my/upload/Inside Investor – Greater KL and Klang Valley.pdf

Perselli, V. (2016). *Education, Theory and Pedagogies of Change in a Global Landscape: Interdisciplinary Perspectives on the Role of Theory in Doctoral Research*. New York: Springer.

Pheng, O. W. (2007). *MS1525:2007: Code of practice on energy efficiency and use of renewable energy for non-residential buildings (First revision)*. Department of Standards Malaysia. Retrieved October 28, 2017, from Retrieved from https://www.scribd.com/doc/61827656/MS1525-2007-Code-of-Pract

Pring, R. (2004). *Philosophy of Educational Research*. London: Continuum.

PWC. (2016). *Technology-enabled manufacturing growth prospects for eastern India*. Confederation of Indian Industry (CII). New Delhi: PWC. Retrieved August 26, 2017, from https://www.pwc.in/assets/

pdfs/publications/2016/technology-enabled-manufacturing-growth-prospects-for-eastern-india.pdf.

Quinlan, C., Zikmund, W., Babin, B., Carr, J., & Griffin, M. (2015). *Business Research Methods* (1st ed.). London: Cengage Learning EMEA.

Roller, M., & Lavrakas, P. (2015). *Applied Qualitative Research Design: A Total Quality Framework Approach* (1st ed.). New York: The Guilford Press.

Roulston, K., & Shelton, S. (2015, April 1). Reconsceptualizing Bias in Teaching Qualitative Research Methods. *SAGE Journals, 21*(4), 332–342. doi:10.1177/1077800414563803

RVO. (2017). *Smart Cities in Malaysia.* Den Haag: Agentchap. Retrieved June 15, 2017, from http://www.rvo.nl/sites/default/files/Smart Cities Malaysia.pdf.

Sadler, M. (2017, July 12). *Can Improved Demand Help Malaysia's Economy Expand in 2Q17?* Retrieved May 5, 2018, from Marketrealist.com Web site: https://marketrealist.com/2017/07/can-improved-demand-help-malaysias-economy-expand-in-2q17

Sahay, A. (2016). *Peeling Saunders's Research Onion.* Retrieved October 11, 2017, from https://www.researchgate.net/publication/309488459_Peeling_Saunder's_Research_Onion

Saldana, J., & Leavy, P. (2011). *Fundamental of Qualitative Research.* (N. Beretves, Ed.) USA: Oxford University Press.

Samsudin, M. S., Rahman, M. M., & Wahid, M. A. (2016). Power Generation Sources in Malaysia: Status and Prospects for Sustainable Development. *Journal of Advanced Review on Scientific Research, 25*(1), 11–28.

Saunders, M., Lewis, P., & Thornhill, A. (2012). *Research Methods for Business Students* (6th ed.). Harlow: Pearson Education Limited.

Schleich, J., & Gruber, E. (2008). Beyond case studies: Barriers to energy efficiency in commerce and the services sector. *Energy Economics, 30*(2), 449–464. doi:10.1016/j.eneco.2006.08.004.

Schock, R., & Sims, R. (2012). *Energy Supply Systems.* International Institute for Applied System Analysis. International Institute for Applied System Analysis. Retrieved August 30, 2017, from http://www.iiasa.ac.at/web/

home/research/Flagship-Projects/Global-Energy-Assessment/GEA_Chapter15_supply_lowres.pdf

Scotland, J. (2012). Exploring the Philosophical Underpinnings of Research: Relating Ontology and Epistemology to the Methodology and Methods of the Scientific, Interpretive, and Critical Research Paradigms. *English Language Teaching, 5*(9), 9–16. doi:doi:10.5

Seai. (2012). *Empowering Businesses, energising Resources: Large industry Energy Network (LIEN) Annual Report 2012.* Sustainable Energy Authority of Ireland (Seai). Retrieved August 28, 2017, from https://www.seai.ie/Your_Business/Large_Energy_Users/LIEN/LIEN_Reports/LIEN-Annual-Report-2012.pdf.

Seido Solutions. (2018). *Energy Efficiency: Air Conditioning Energy Saver.* Retrieved January 9, 2018, from Seido Soutions Web Site: https://www.seidosolutions.com/energy-efficiency/led-lighting/

Seido Solutions. (2018). *Energy Efficiency: Led lighting Solutions.* Retrieved January 9, 2018, from Seido Solutions Web Site: https://www.seidosolutions.com/energy-efficiency/led-lighting/

Silverman, C. (2016). *Construction Term of the Month: Building Envelope.* Retrieved May 4, 2018, from silvermancpm.com Web Site: http://silvermancpm.com/construction-term-of-the-week-building-envelope/

Silverman, D. (2015). *Interpreting Qualitative Data* (5th ed.). Melbourne. Retrieved June 10, 2017, from https://books.google.co.in/books?id=BvmICwAAQBAJ&dq=Rhetorical+assumption+qualitative+research&lr=&source=gbs_navlinks_s

Silvius, A., & Schipper, R. (2009). *A Maturity Model for Integrating Sustainability in Projects and Project Management.* Retrieved October 25, 2017, from Researchgate.net Web site: https://www.researchgate.net/publication/267228611_A_Maturity_Model_for_Integrating_Sustainability_in_Projects_and_Project_Management.

Silvius, A., Brink, J. V., & Köhler, A. (2010). The impact of sustainability on Project Management. *Asia Pacific Research Conference on Project Management (APRPM).* Melbourne: APRPM.

Sindhu, A. (2012). *Sales promotion strategy of selected companies of FMCG Sector in Gujarat region.* Retrieved October 27, 2017, from shodhganga.inflibnet: http://shodhganga.inflibnet.ac.in/handle/10603/3704/.

Smith, D. (2016). *Phenomenology, The Standard Encuclopedia of Philisophy.* Standford: Metaphyscial Research Lab.

Sorrell, S., Mallett, A., & Nye, S. (2011). *Barriers to industrial energy efficiency: A literature review.* United Nations Industrial Development Organisation (UNIDO), Development, Policy, Statistics and Research Branch, Vienna. Retrieved August 7, 2017, from http://www.unido.org//fileadmin/user_media/Publica...

Srinivas, S., Gadde, B., Seth, S., & Dhage, V. (2015). Implementing Energy Efficiency in Buildings: A compendium of experiences from across the world. *International Conference on Energy Efficiency in Building (ICEEB 2015), Delhi.* Delhi: United Nations Development Programme (UNDP). Retrieved November 20, 2017, from http://www.undp.org/content/dam/india/docs/ICEEB 2015_Compendium.pdf.

Sun, T. (2009). *Mixed methods research: Strengths of two methods combined.* SMC University, Management. Zug: SMC University.

Sussex. (2017). *Understanding barriers to energy efficiency.* Retrieved May 26, 2017, from Univeirsity of Sussex Web site: http://www.sussex.ac.uk/Units/spru/publications/reports/barriers/finalsection3.pdf

Tan, C., & Nikkei. (2017). *Malaysia's exports hit record high in March.* Retrieved May 4, 2017, from Asia Nikkei Web site: https://asia.nikkei.com/Economy/Malaysia-s-exports-hit-record-high-in-March

Tan, Y., Shen, L., & Yao, H. (2011). Sustainable construction practice and contractors' competitiveness: a preliminary study. *Habitat International,* 225–230.

Tanaka, K., Watanabe, H., & Endou, A. (2010). Enerize E3 Factory Energy Management System. *Yokogawa Technical Report English Edition,* 23–26. Retrieved June 15, 2017, from Tanaka, K., Watanabe, H., & Endou, A. (2010). Enerize E3 Factory Energy Management System. Yokogawa Technical Report English Edition, 53(1), 23–26. Retrieved from http://web-material3.yokogawa.com/rd-te-r05301-005.pdf.: http://web-material3.yokogawa.com/rd-te-r05301-005.pdf.

Taylor, J. (2008). Organizational Influences, Public Service Motivation and Work Outcomes. *International Public Management Journal, 11*(1), 67–88.

TERI. (2013). *Energy Efficiency and Renewable Energy in Leading Indian Corporates.* The Energy and Resources Institute. New Delhi: The Energy and Resources Institute (TERI). Retrieved June 23, 2017, from http://cbs.teriin.org/pdf/reports/Energy_Efficiency_Compendium.pdf.

Tey, N. P. (2014). Inter-state Migration and Socio-demographic Changes in Malaysia. *Malaysian Journal of Economic Studies, 51*(1), 121–139. Retrieved June 23, 2017, from https://umexpert.um.edu.my/file/publication/00001678_106796.pdf.

Thanh, N. C., & Thanh, T. T. (2015). The Interconnection Between Interpretivist Paradigm and Qualitative Methods in Education. *American Journal of Educational Science,, 1*(2), 24–27. Retrieved May 8, 2018, from https://pdfs.semanticscholar.org/79e6/888e672cf2acf8afe

The Need Project. (2017). *Energy Consumption.* Retrieved May 6, 2018, from need.org Web site: http://www.need.org/Files/curriculum/infobook/ConsI.pdf

Thollander, P., Sa, A., Paramonova, S., & Cagno, E. (2015, August). Classification of Industrial Energy Management Practices: A Case Study of Swedish Foundary. *Energy Procedia, 75,* 2581–2588. doi:10.1016/j.egypro.2015.07.311

Thomas White International. (2014). *Manufacturing Drives Growth.* Retrieved May 4, 2018, from Thomas White International Web Site: https://www.thomaswhite.com/world-markets/malaysia-manufacturing-drives-growth/

Thomas, G. P. (2012). *What is Climate Proofing?* Retrieved October 11, 2017, from Azocleantech Web site: https://www.azocleantech.com/article.aspx?ArticleID=268

Thomson, S. (2011). Sample size and grounded theory. *Journal of Administration & Governance, 5*(1), 45–52.

Tiesen, R. (2011). *After Godël: Platonism and Rationalism in Mathematics and Logic.* Oxford and New York: Oxford University Press.

Tremblay, K., Lalancette, D., & Roseveare, D. (2012). *Assessment of Higher Education Learning Outcomes: Feasibility Study Report.* Retrieved

August 28, 2017, from Organization for Economic Co-Operation and Development (OECD) Web site: http://www.oecd.org/education/skills-beyond-school/AHELOFSReportVolume1.pdf.

Turner, J. (2010). *Responsibilities for Sustainable Development in Project and Program Management. Expert Seminar Survival and Sustainability as Challenges for Projects.* Zurich: International Project Management Association (IPMA).

UNEP. (2009). *A Guide to Greenhouse Gas Emission Reduction in UN Organizations.* Retrieved August 15, 2017, from UNEP Web site: Retrieved from http://www.greeningtheblue.org/sites/default/files/EmissionReductionGuide.pdf.

UNEP. (2015). *Accelerating Energy Efficiency: Initiatives And Opportunities – Southeast Asia.* United Nations Environment Program (UNEP). UNEP. Retrieved July 14, 2017, from http://www.fundacionbariloche.org.ar/wp-content/uploads/2015/11/C2E2_REPORT_LAC.pdf.

UNEP; BCA. (2011). *Sustainable Building Policies on Energy Efficiency,Country Report on Sustainable Building Policies on Energy Efficiency in Brunei Darussalam, Cambodia, Indonesia, Malaysia, Myanmar, Philippines, Singapore, Thailand, Vietnam. Singapore:.* Retrieved January 9, 2018, from BCA Web site: http://www.csb.sg/regional-study.html

UNFCCC. (1998). *Kyoto Protocol to the United Nations Framework Convention on Climate Change.* Bonn: United Nations Framework Convention on Climate Change (UNFCCC). Retrieved June 2, 2017, from www://unfcc.int/resource/docs/convkp/kpeng.pdf

UNFCCC. (2006). *Clean Development Mechanism.* Bonn: United Nations Framework for Convention for Climate Change (UNFCCC). Retrieved June 2, 2017, from www://unfcc.int/kyoto_protocol/mechanism/clean_development_mechanism/item/2718.php

UNFCCC. (2012). *Doha Amendment to the Kyoto Protocol tothe United Nations Framework Convention on Climate Change.* Amendments, United Nations Framework Convention on Climate Change (UNFCC), Bonn. Retrieved June 2, 2017, from http://unfccc.int/files/kyoto_protocol/doha_amendment/application/pdf/attachment_sg_letter_doha_amendment.pdf

UNICEF. (2017). *Overview: Data Collection and Analysis Methods in Impact Evaluation*. United Nations Children Fund (UNICEF), New York. Retrieved August 26, 2017, from https://www.unicef-irc.org/publications/pdf/brief_10_data_collection_analysis_eng.pdf.

UNIDO. (2017). *Module 12: Energy efficiency technologies and benefits*. Retrieved July 13, 2017, from UNIDO Web site: https://www.unido.org/fileadmin/media/documents/pdf/EEU_Training_Package/Module12.pdf.

United Nations Economic Commission for Europe. (2017). *Best Policy Practices for Promoting Energy Efficiency*. Retrieved August 30, 2017, from United Nations Web site: http://www.un-expo.org/wp-content/uploads/2017/06/ECE_Best_Practices_in_EE_publication.pdf.

Vosloo, J. J. (2014). *A sport management programme for educator training in accordance with the diverse needs of South African schools*. North-West University. Retrieved December 15, 2017, from https://dspace.nwu.ac.za/handle/10394/12269?show=full

Wang, J. (2014). *Encyclopedia of Business Analytics and Optimization*. USA: IGI Global.

Wang, J., Li, Z., & Tam, V. W. (2014). Critical factors in effective construction waste minimization at the design stage: a Shenzhen case study. *China. Resour. Conserv. Recycl, 82*, pp. 1–7.

Wellington, J. (2000). *Educational Research: Contemporary Issues and Practical Approaches*. London: Continuum.

White, T. I. (1993). *Business Ethics: A Philosphical Reader*. New York: Macmillian.

Wong, P. S., Ng, S. T., & Shahidi, M. (2013). Towards understanding the contractor's response to carbon reduction policies in the construction projects. *International Journal of Project Management, 31*, 1042–1056.

World Bank. (2011). *The World Bank Annual Report 2011 : Year in Review*. World Bank, Washington D.C. Retrieved June 20, 2017, from World Bank Web Site: https://openknowledge.worldbank.org/handle/10986/2378

World Bank. (2016). *Annual report 2016.. Retrieved from*. Annual Report, Washington (DC). Retrieved May 4, 2018, from World Bank Web site: https://openknowledge.worldbank.org/handle/10986/24985

World Bank Group. (2016). *Doing Business 2016 Measuring Regulatory Quality and Efficiency. Washington.* World Bank. World Bank. Retrieved August 20, 2017, from http://www.doingbusiness.org/~/media/WBG/DoingBusiness/Documents/Annual-Reports/English/DB16-Full-Report.pdf.

World Economic Forum. (2017). *The Future of Electricity New Technologies Transforming the Grid Edge.* Retrieved July 25, 2017, from World Economic Forum Web site: http://www.weforum.org/docs/WEF_Future_of_Electricity_2017.pdf.

World Energy Council. (2013). *World Energy Perspective Energy Efficiency Technologies. London.* London: World Energy Council. Retrieved August 26, 2017, from Retrieved from https://www.worldenergy.org/wp-content/uploads/2014/03/World-Energy-Perspectives-Energy-Efficiency-Technologies-Overview-report.pdf

Worrell, E. (2011). *Barriers to energy efficiency: International case studies on successful barrier removal (Working Paper No. 14/2011).* United Nations Industrial Development Organization (UNIDO), UNIDO. Retrieved September 14, 2017, from https://www.unido.org/fileadmin/user_media/Services/Research_and_Statistics/WP142011_Ebook.pdf.

Worrell, E., Martin, N., Price, L., Ruth, M., Elliott, N., Shipley, A., & Thorn, J. (2003). *Emerging Energy-Efficient Technologies for Industry.* Washington D.C: International Energy Studies Group. Retrieved December 21, 2017, from Retrieved from https://ies.lbl.gov/sites/all/files/lbnl-50527.pdf.

Worrell, E., Van Berkel, R., Fengqi, Z., Menke, C., Schaeffer, R., & O. Williams, R. (2001). Technology transfer of energy efficient technologies in industry: a review of trends and policy issues. *Energy Policy,* 29–43. doi:10.1016/S0301-4215(00)00

Wright, T., Klein, M., & Wieczorek, J. (2015). *Ranking Populations Based on Sample Survey Data.* Pittsburg: Carnegie Mellon University.

Xu, M., & Storr, G. (2012). Learning the Concept of Researcher as Instrument in Qualitative Research. *The Qualitative Report, 17*(21), 1–18. Retrieved from https://nsuworks.edu/tqr/vol17/iss21/2

Yan, M. R., & Chien, K.-M. (2013). Evaluating the Economic Performance of High-Technology Industry and Energy Efficiency: A Case Study of Science Parks in Taiwan. *Energies, 6*(2), 973–987. doi:10.3390/en6020973.

Yin, R. K. (2009). *Case Study Research: Design and Methods* (4th ed.). Thousand Oaks: SAGE Publications.

Zaid, S. M., Myeda, N. E., Mahyuddin, N., & Sulaiman, R. (2014). Lack of Energy Efficiency Legislation in the Malaysian Building Sector Contributes to Malaysia's Growing GHG Emissions. *E3S Web of Conferences, 3*, 1029. doi:10.1051/e3sconf/20140301029

Zaid, S. M., Myeda, N. E., Mahyuddin, N., & Sulaiman, R. (2015). Malaysia's Rising GHG Emissions and Carbon' Lock – In' Risk: A Review of Malaysian Building Sector Legislation and Policy. *Journal of Surveying, Construction and Property (JSCP), 6*(1), 1-13.

Zhu, Q., & Geng, Y. (2013). Drivers and barriers of extended supply chain practices for energy saving and emission reduction among Chinese manufacturers. *Journal of Cleaner Production, 40*, 6–12. doi:10.1016/j.jclepro.2010.09.017.

Zulkifli, Z. (2016). *Malaysia Country Report. In S. Kimura & P. Han (Eds.), Energy Outlook and Energy Saving Potential in East Asia 2016. ERIA Research Project Report 2015-5*. Economic Research Institute for ASEAN and East Asia (ERIA), Jakarta.

Appendices

Appendix A

Consent (for e-mail)

Dear Participant,

I am a doctoral student at SMC University pursuing a Doctor of Management degree. I am conducting dissertation research to understand "Perception of energy experts on commercial and industrial electricity consumers (CIEC) of Klang Valley, Malaysia toward the adoption of energy-efficient technologies." Considering the above primary research objective, the research aims to learn about the attitudes of consumers of electricity toward the adoption of energy-efficient technology from the viewpoint of energy experts in the commercial and industrial sectors in the Klang Valley. I intend to confine this research to energy experts involved in providing energy advisory, audit, and management services to commercial and industrial facilities within Klang Valley, Malaysia.

I am inviting you to take part, at your convenience, in this important survey which will take approximately 45–60 minutes of your time. I would like to meet with you personally and conduct a face-to-face interview at a location convenient to you and, with your consent, record the responses to the questions for each interview focused on your experiences.

Since it is voluntary for you to participate in this study, your relationship with me or SMC University will be kept confidential. This study is anonymous, and we will not be publishing in the dissertation any information about your identity.

Feel free to contact me if you have any further questions relating to the study, Thirumalaichelvam Subramaniam at thirumalaichelvam.subramaniam@student.swissmc.ch or by telephone at +60 12 297 6799. Upon the completion of the interview and within a couple of days, a copy of the interview transcript will be emailed to you for you to review the accuracy of the content, and if there are any changes, you are required to e-mail the amended transcript to the undersigned as soon as possible. As a research participant, if you have any other concerns about your rights that have not been answered by me, you may also contact my mentor Dr. Jeffrey Ray at j.ray@smcuniversity.com.

Thank you in advance for your cooperation in this matter.

Ir. Thiru S.
Thirumalaichelvam Subramaniam
Doctoral Candidate
Swiss Management University

Appendix B
Individual Consent

STATEMENT OF CONSENT FOR INTERVIEW

A copy of this consent form will be provided to you for your records.

In signing this consent, I acknowledge that:

- I have read and understood the information provided to me regarding this study, and I am willing to participate in this research to understand "Perception of energy experts on commercial and industrial electricity consumers (CIEC) of Klang Valley, Malaysia toward the adoption of energy-efficient technologies."

- I understand that I will be asked a number of questions in a face-to-face interview with the researcher to capture my livid experience on the primary topic mentioned above, which will take 45–60 minute, and I have given my permission to digitally audio record this interview.

- I understand that my participation in the interview process is voluntary and may withdraw at any point of time during this research study. I may not provide any reason for leaving the study. My relationship with the researcher or Swiss Management Centre University will not be affected in case I may refuse to take part in the study at any time.

- I understand that at any point of time during the process, I have the right not to answer any single question, as well as to withdraw completely from the interview, and I have the right to request that the interviewer not use any of my interview material.

- I am advised that there is no direct benefit or compensation for my participation in this study.

- This study is anonymous, and we will not be collecting or retaining any information about your identity. The records of this study will be kept strictly confidential. Research records will be kept in a locked file, and all electronic information will be coded and secured using a password protected file, including audio or video recordings made during the interviews. The researcher and the researcher's mentor will be the only persons having access to the research records. The research or study records and audio or video recording will be kept for 5 years from the first date of recording. Research records will be destroyed by method of mechanical shredding, and audio and video storage will be destroyed by means of incineration after the expiry of the 5-year period mentioned above. In order to protect your identity, we will not include any information in any report we may publish that would identify you.

I hereby declare that all aforementioned statements have been read, and I have understood all the statements. I agree to the terms and conditions of the consent, and my participation will be entirely voluntarily during the project.

Name of the participant:	Name of the Researcher:
Date:	Date:
Place:	Place:
Signature:	Signature:

Appendix C
Interview Protocol

'Good morning/afternoon/evening, (name of participant). First of all, I would like to welcome you and thank you for your time in supporting me by participating in this dissertation research study.

My name is Thirumalaichelvam Subramaniam, and I have undertaken studies for the Doctor of Management program as a post-graduate student at Swiss Management Center University.

This dissertation study is being completed to partial fulfillment of the requirements for this doctoral degree. This interview with you for this research study will take approximately 45 to 60 minutes of your time. I intend to interview you by asking a number of questions to capture your experiences about "Perception of energy experts on commercial and industrial electricity consumers (CIEC) of Klang Valley, Malaysia toward the adoption of energy-efficient technologies." I wish to record this interview and therefore seek your permission, so I can accurately document the information that you will be providing to me. Do I have your permission to do so?

Please feel free to inform me if, at any time during the interview, you do not wish to proceed with this interview or recording. The responses you will provide to the questions will enable me to gain further understanding of your perceptions on the primary research question that I stated earlier. I guarantee that the responses you provide will only be used for my dissertation research report and will be kept strictly confidential and anonymous.

I would, therefore, now like you to sign the consent document to confirm your agreement to participate in this study. I will provide you one copy to keep, and I will securely store the other copy and keep it confidential, separate from your reported responses. Thank you.

A. PRELIMINARY QUESTIONS

1. What is your profession?

    1b. What are your academic and professional qualifications?

2. How are you employed in the company you are working in currently?

3. How long have you been with this company, or if this is your own business, how long has it been operating?

4. How many people, including you, work in this business?

5. What is your designation in the company?

6. How long have you been involved in your profession?

7. How long have you been involved in the field of energy efficiency (EE)?

8. How are you involved with Commercial and Industrial Electricity Consumers (CIEC) in Klang Valley, Malaysia?

9. What are the types of services you provide to CIEC of the Klang?

B. CORE DISSERTATION RESEARCH QUESTIONS

Questions asked during the interview will be focused on the topic of energy efficiency based on the relevant research questions such as:

B-1 RQ#1. Why Are Electricity Consumers in Klang Valley Resistant to Adopting Energy-Efficient Postures?

Question based on Energy Experts' perceptions of CIEC of Klang Valley, Malaysia.

B-1-1 Questions on Policy and Regulatory:

1. How is the cooperation between local agencies and organizations responsible for EE implementation?

2. What support in terms of meeting policy and regulatory requirements in EE is provided to your CIEC of the Klang Valley clients by local agencies or organizations responsible for EE?

3. What are the performance indicators on energy consumption established by your CIEC of the Klang Valley clients?

4. What are the permits related to environment and technical (safety) required when implementing energy-efficient project for your CIEC of Klang Valley clients? Describe its strictness.

5. Would your CIEC of the Klang Valley clients accept EE measures recommendations based on regulatory evaluation, and why?

6. How does the control imposed by the electricity market affect your CIEC of the Klang Valley clients in implementing EE technologies?

7. Are there any strict regulations imposed by the government or its agency on energy efficiency implementation on your CIEC of the Klang Valley clients? If yes, how are the regulations enforced?

8. What are the incentives provided by the government on implementation of EE technologies for your CIEC of the Klang Valley clients?

9. Can government regulations and policies be easily implemented by your CIEC of the Klang Valley clients? How?

10. What are the infrastructures in place for EE that your CIEC of the Klang Valley clients can easily adopt?

11. Would industrial regulation in utility usage encourage your CIEC of Klang Valley clients to implement EE measures? Why?

12. What internal policies on energy management are implemented by your CIEC of Klang Valley clients?

13. What type of matured and proven EE technologies have been implemented on EE projects by your CIEC of the Klang Valley clients?

14. What energy management policies have been incorporated into your CIEC of the Klang Valley client's organizational policies?

15. What kind of environmental policies in regard to the reduction of GHG or CO_2 emissions have been implemented in your CIEC of the Klang Valley client's organizations?

B-1-2 Questions on Economic, Finance, and Marketing:

1. What are the studies made in regard to the reduction in EE gap based on the existing technology being utilized and the currently available technology?
2. What is the demand for EE project implementation in the Klang Valley region of Malaysia?
3. How is the supply of EE equipment or system?
4. What is an accurate method to determine the financial success of any EE project carried out by your CIEC of the Klang Valley clients?
5. What is the payback period expected by your CIEC of the Klang Valley clients for EE projects?
6. What are the missed financial opportunities experienced by the CIEC of the Klang Valley clients for EE measures, or would they rather focus on improving their core business processes?
7. What type of EE project gets financial recognition among your CIEC of the Klang Valley?
8. Since Malaysia has one of the lowest electricity tariffs among the ASEAN countries, do you believe that this poses a disincentive to your CIEC of the Klang Valley clients to implement energy-efficient measures?
9. What type of finances are made available to your CIEC of the Klang Valley clients for EE training for their in-house personnel?
10. What type of your CIEC of the Klang Valley clients is able to afford in-house energy personnel?
11. Is EE project given priority over other projects in your CIEC of the Klang Valley clients' organization? If no, why?
12. What are the financial risks that your CIEC of the Klang Valley clients might face when EE projects are implemented?
13. How can your CIEC of the Klang Valley clients access internal or external funds for EE projects?
14. What are the effects budget policies limit have on the CIEC of the Klang Valley EE projects?

15. What kind of internal policy is in place in your CIEC of the Klang Valley client's organization on cost reductions?

B-1-3 Questions on Behavioral, Information, and Technical:

1. How can your CIEC of the Klang Valley clients' in-house personnel gain knowledge on the available EE technologies?

2. Describe the type of EE training necessary for your CIEC of the Klang Valley client's in-house personnel.

3. Is there a sufficient number of installation personnel for EE equipment in the Klang Valley Malaysia?

4. What standard energy management systems have been implemented by your CIEC of the Klang Valley clients?

5. Describe the awareness of EE among the top management of your CIEC of the Klang Valley clients.

6. Describe the awareness of EE among the employees of your CIEC of the Klang Valley clients.

7. How is energy management system (EnMS) integrated by your CIEC of the Klang Valley clients into their overall management system?

8. How is information gathered by your CIEC of the Klang Valley clients' organization in regard to their energy consumption trends?

9. How can your CIEC of the Klang Valley clients gain information about the available EE technologies?

B-2 RQ#2. How Can Energy Efficiency Technologies be Implemented in the Klang Valley in Malaysia?

1. What are the guidelines provided for EE implementation by local government agencies responsible for EE to your CIEC of Klang Valley clients?

2. What type of existing and proven EE models have been implemented at your CIEC of Klang Valley clients?

3. How is artificial intelligence being utilized for EE measures by your CIEC Klang Valley Clients?

4. What type of proven policies and technology are utilized to reduce GHG and CO_2 among your CIEC of the Klang Valley clients?

5. How are your CIEC of the Klang Valley clients' facilities being certified as being energy efficient?

6. How is the Clean Development Mechanism (CDM) process implemented by your CIEC of the Klang Valley clients?

7. How are energy accounting, energy audits or baselines established for areas of higher energy use, wasted energy, energy savings potential, and energy consumption patterns by your CIEC of the Klang Valley clients?

8. What are the energy management documented procedures which are consistent and written for easy understanding, which are distributed across the entire organization with the necessary resources, available at your CIEC of the Klang Valley clients?

9. What are the effective energy management processes implemented to adjust and optimize energy demands by utilizing EE technologies and procedures for your CIEC of the Klang Valley clients?

10. How is energy maturity matrix, which provides high-level assessment of strength and weaknesses across six areas of energy management, which are energy policy, organizing, training, performance measurement, communication, and investment, being utilized by your CIEC of the Klang Valley clients?

11. What energy management systems (EnMS), such as ISO50001, have been implemented by your CIEC of the Klang Valley clients?

12. What are the Best Available Technologies (BATs) implemented at your CIEC of the Klang Valley clients' facilities?

13. How is your CIEC Klang Valley clients' facilities equipment being monitored?

14. What kind of online execution system present to the end-user via user interfaces, which helps monitor and control energy usages, is being utilized by your CIEC Klang Valley clients?

15. How do your CIEC of the Klang Valley clients collaborate with any known organization to implement EE action plans and EE programs?

16. How do your CIEC of the Klang Valley clients scale-up investment due to policy driven by the local government or by demand driven by Energy Services Companies (ESCOs)?

B-3 RQ#3. How Do Consumers Find Energy-Efficient Technology Useful?

1. How can the implementation of EE measures by your CIEC of the Klang Valley clients significantly reduce the negative impact of energy use on the environment and human well-being, and increase the availability of primary energy reserves while achieving maximum benefits in terms of output from the available energy?

2. What kind of economic gains can your CIEC of the Klang Valley clients achieve by implementing EE measures or technologies?

3. How can your CIEC of the Klang Valley clients be cost-effective to achieve energy security, to improve business productivity, and to mitigate GHG and CO_2?

4. How can EE measures make good business sense for your CIEC of the Klang Valley clients?

5. How can the rising fuel prices augment savings in the long run for your CIEC of the Klang Valley clients?

6. How can EE implementation help your CIEC of the Klang Valley clients survive economic downturn compared to those who have not implemented EE?

7. How can the implementation of EE technologies help your CIEC of the Klang Valley clients reduce energy cost when they undergo a business expansion which requires more energy consumption?

8. How can your CIEC of the Klang Valley clients contribute to the reduction of local CO_2 and GHG emission and be socially responsible (CSR)?

Appendix D
Interviewee's Transcript Review (ITR) (for e-mail)

Dear Participant,

First, I would like to thank you for participating in the recent interview for my doctoral dissertation research on the "Perception of energy experts on commercial and industrial electricity consumers (CIEC) of Klang Valley, Malaysia toward the adoption of energy-efficient technologies."

Attached is your interview transcript document, which is a summary of interim results developed through analysis of your interview to represent your perception on the research topic. Please read the document and comment on whether or not you felt the synthesized results resonated with your experiences and if there is anything you would like to change, in order to help us complete our analyses and develop interpretations.

Appreciate if you could return the transcript with changes (if any) to the undersigned by e-mail as soon as possible upon the completion of your review.

Feel free to contact me if you have any further questions relating to the study, Thirumalaichelvam Subramaniam at thirumalaichelvam.subramaniam@student.swissmc.ch or by telephone at +60 12 297 6799. As a research participant, if you have any other concerns about your rights that have not been answered by me, you may also contact my mentor Dr. Jeffrey Ray at j.ray@smcuniversity.com.

Thank you in advance for your cooperation in this matter.

Ir. Thiru S.
Thirumalaichelvam Subramaniam
Doctoral Candidate
Swiss Management University

www.ingramcontent.com/pod-product-compliance
Lightning Source LLC
Chambersburg PA
CBHW020859180526
45163CB00007B/2561